Agriculture and Food Production

Agriculture and Food Production

Dr. B.K. Sharma

AMIGA PRESS. INC
Delhi -110051

AMIGA PRESS. INC

H.O :- 28-A, Indra Park, Chander Nagar

Delhi :- 110051 (India)

Tel :- 011-22504028, +91-98736-57000

Email :- amigapress@gmail.com

amigaeditorial@gmail.com

amigapub@gmail.com

Website :- www.amigapress@co.in

This edition is published by Amiga Press.Inc

ISBN : 978-93-84533-36-6

First Published 2018

Printed and bound by Replika Press Pvt. Ltd.

Preface

Agriculture plays a valuable role in our everyday lives by not only providing us with food, but also by maintaining a strong economy. On a worldwide basis, more people are in some way involved in agriculture than in all other occupations combined.

Developments in science and technology have contributed to better soil, nutrient, water, and pest management, and to more efficient methods of planting, harvesting, storing, processing, and transporting farm products. These developments have resulted in more and better quality food than ever before.

To keep agriculture on the cutting edge, many farmers now implement a variety of technologies, including precision agriculture, remote sensing, computers, the internet, specialized software, global positioning, drip irrigation, and biotechnology. The Global Positioning System (GPS) is a key technology utilized in precision agriculture. Linking to a system of satellites, a farmer uses a receiver to pinpoint his or her position to within inches. The information helps the farmer identify precisely where to plant, and when and where to apply pesticides and fertilizer.

Modern biotechnology includes a range of tools, including genetic engineering, that are utilized to develop beneficial traits in plant and animal agricultural products. For example, crops such as corn and soybean have been genetically enhanced for improved weed, pest, and disease management, reduced pesticide use, higher-yielding crops, reduced soil erosion, and reduced levels of natural toxins.

Meeting the food needs of a still-growing and more affluent global population—as well as the nearly one billion people who already go without adequate food—presents a key challenge for economic and human security. Many analysts estimate that food production will need to nearly double over the coming several decades. Recent trends of using food crops for fuel (e.g., corn ethanol) or displacing food crops with fuel crops, along with potential opportunities for reforesting land for carbon credits, may amplify the food security challenge by increasing competition for arable land. Climate change increases the complexity of meeting these food needs because of its multiple impacts on agricultural crops, livestock, and fisheries. The potential ability of agricultural and fishery systems to limit climate change adds yet another dimension to be considered.

Biotechnology is necessary to maintain our agriculture competitive and remunerative and to achieve nutrition security in the face of major challenges such as declining per capita availability of arable land; lower productivity of crops, livestock and fisheries, heavy production losses due to biotic (insects pests, weeds) and abiotic (salinity, drought, alkalinity) stresses; heavy post-harvest crop damage and declining availability of water as an agricultural input. Investment in agricultural related biotechnology has resulted in significantly enhanced R&D capability and institutional building over the years. However, progress has been rather slow in converting the research leads into usable products.

This book covers all the aspects of agriculture. It is hoped, the book will be found to be of immense value to the students of higher education, researchers and teachers.

—Editor

Contents

1

Food Production

THE BASICS OF FOOD

Your body's goal is to digest food and use it to keep your body alive. In the following sections, we will look at each of these basic components to understand what they really do and why they are so important to your body.

General Supplements

Antioxidants

Vitamins, minerals, and other molecules, which act in the body to scavenge oxygen radicals, are antioxidants. Most patients that need nutritional support will benefit from a high level of dietary antioxidants. Antioxidants should be supplemented as a group instead of high levels of individual ones, due to the fact that any antioxidant can act as a pro-oxidant if present in high enough amounts.

Minerals

Minerals act as cofactors for enzymes for almost every reaction in the body. Everything from immune system function, to bone density, to protein, fat and CHO metabolism is affected by mineral deficiencies. Mineral supplementation should be critically evaluated.

The authors suggest amino acid chelated minerals, due to the increased bioavailability of these formulations. Other organic complexes (citrates, gluconates, and lactates) have a higher biological value than the inorganic complexes (oxides, carbonates) with sulfates being in the middle. The inorganic forms of many minerals are toxic, while the organic forms are very beneficial (ex: chromium). Unless minerals are supplied in the correct balance and form, minimal benefit will be obtained by supplementation.

Vitamins

Vitamins are categorized into two distinct classes: water soluble and fat soluble. Water soluble vitamins are not stored in the body; they are excreted if not utilized soon after ingestion. Fat soluble vitamins (Vitamins A, D, E, and K) are stored and have more of a potential for toxicity.

Horses on premium commercial feeds rarely develop a clinical deficiency of an individual mineral or vitamin because the feeds are appropriately fortified. General supplements should not be used to try to make a good ration out of cheap feedstuffs. They can be used to support horses that are under more stressful conditions such as performance, reproductively active, diseased, or geriatric.

Hoof

Hoof health is often a reflection of the nutritional status of the individual. Keratin based structures such as the hoof, mane, tail and hair may reflect nutritional deficiencies long before other body systems begin to exert obvious signs. Vitamins, minerals, and specific amino acids maybe spared incorporation into these structures due to the need in more biologically "important" structures. However some individuals may need supplementation for the hoof and other keratin based structures beyond what can be supplied in even premium horse feeds. In these cases attention should be given to specific nutrients that are necessary for proper keratin formation. Guidelines provided in this chapter for evaluation of any supplement should be applied when choosing a "hoof supplement". In addition the following is a list and short description of specific nutrients that provide the body with the basic building blocks of keratin.

Biotin: Biotin is a B vitamin that is required by all animals as a biological catalyst in numerous metabolic reactions. Biotin is involved in the keratinization process early on in the horn cell's development. Due to its early incorporation, the benefit of biotin supplementation is often not seen for months or until new hoof growth is evident. Animals with biotin deficiency present with dry and cracked skin, poor hair coats, split and cracked hooves and foot lesions. Though biotin may be the most well known "hoof nutrient" it is not the only one. Some horses supplemented with biotin alone exhibit no improvement in overall hoof health. This often suggest a deficiency in other nutrients that are integral in healthy hoof development.

Amino Acids: Lysine and Methionine are two amino acids essential for quality hoof growth. Keratin is a protein. Proteins are huge complex molecules that involve numerous amino acids for the complete formation. Proteins that lack enough of the correct types of amino acids often are not functionally sound, resulting in poor quality hooves. Methionine and Lysine are highly utilized in the horse and should be supplied in adequate amounts in the diet.

Joint

It was found over 50 years ago, that over time, eating ground up connective tissues could help relieve problems associated with arthritis. People began to consume gelatin and cartilage (shark, chicken, bovine, or Perna muscle) and these helped some. Studies found that chondroitin sulfates were contributing most of the beneficial effects. About 30 years ago the research began focusing in on the much smaller Glucosamine Sulphate.

Whole cartilage and its large components (chondroitin sulfates and collagen) can not be absorbed well by the body. While large molecules such as these can be injected and work very well, when taken orally more than half usually ends up in the manure. Chondroitin sulfates have a large range in size. Some are 50 times larger than others. Only 8-10% of the smallest ones have been shown to get from the gut into the blood intact. What is fed must be digested by the gut and the pieces that do get into the blood must be broken down further to get into the joints.

There are several types of glucosamine used in supplements. Glucosamine Sulphate is the basic building block of connective tissues and fluids. Studies have shown in humans, rats, and dogs, that when Glucosamine Sulphate is given orally, within 30 minutes 87-97% is actively taken from the gut into the blood. Further, within 4 hours the chondrocytes (joint cells) have actively taken it from the blood. When linked together with sulfur, chains of glucosamines become several different types of connective tissues and joint fluid. [A lack of sulfur will cause the production of connective tissues to stop.] Also an enzyme can slightly change the shape of the glucosamine into galactosamine, the basic building block of the chondroitin sulfates. In one clinical study, where Glucosamine Sulphate was feed to horses, 77% of Navicular disease and 100% of Spavin cases returned to normal function.

N-acetyl-glucosamine has been proven to not have active uptake from the gut in vitro. There are no studies that show if Glucosamine HCl is actively absorbed, or how much of it actually gets into the blood or the joints. Whatever portion does make it into the joints must be undergo changes before it can be used in connective tissues. The HCl must be removed and a sulphate added. Furthermore, Glucosamine HCl is not stable in liquids. Published experiments have shown that over half of the Glucosamine HCl added to a liquid solution will breakdown within 27 hours. Within 4 days, these new and different "breakdown products" form completely different molecules that were 20 times larger than what was originally in the solution.

Other non-structural ingredients are also commonly used. Glutamic acid (glutamate) is an excitatory amino acid (EAA). When this amino acid is added to a brain experimentally it causes seizures. Higher than normal concentrations are present in the brain of human seizure patients. Research is being done to see whether blocking its activity will help control seizures. It is also being considered as an antidepressant drug. This amino acid is also naturally released by joint nerves to signal pain. It is used as an experimental marker to determine the amount of pain induced in experimental models. The more glutamate that is released, the more pain that is felt in a joint. Caution is given here because a review of this product in a popular publication discusses its dramatic and swift change in horses' attitudes. Research is currently being done to determine its role in seizures and the signalling of pain. What will prolonged use do? At what levels does this ingredient become dangerous? How addicting is it? These are only some of the questions not clearly answered by science. Glutamic acid is added to some joint supplements because the amine group in glucosamine can be donated by glutamic acid. However, providing extra of a particular AA does not mean that more connective tissue will be made.

Bromelain is an enzyme found in pineapples. It has been shown to reduce the amount of swelling present in some experimental animals. However, Bromelain can enter the body intact, because it digests the lining of the GI tract. In human cancer patients, it has been shown to stimulate the body's own immune system to kill somatic cells. Since it is a protein, it can cause allergic reactions. People who handle bromelain have become allergic to it.

Many of the "flex products" on the market contain herbs, which do not contribute nutrients to the joint tissues. While an in depth discussion of herbal ingredients is beyond the scope of this paper, the following are mentioned due to their common addition to many supplements and their serious impact on or conflict with other drugs and their potential to cause positive results on competitive drug screenings.

Yucca contains steroid saponins. Steroids have been shown to slow the production of glycosaminoglycans (GAG's). Boswellia, white willow's bark, and snake root are all herbs containing chemicals that act as NSAID's. Devil's claw contains several chemicals, which are reported to decrease pain. They do not decrease inflammation and are similar in structure to steroids. This herb should not be used with any female (human or animal), as it has been reported that Devil's claw can cause abortions, by stimulating uterine contractions.

FOOD PRODUCTS

Activities of food scientists include the development of new food products, design of processes to produce these foods, choice of packaging materials, shelf-life studies, sensory evaluation of the product with trained expert panels or potential consumers, as well as microbiological and chemical testing. Food scientists at universities may study more fundamental phenomena that are directly linked to the production of a particular food product and its properties. In the U.S., food science is typically studied at land-grant universities.

Food science is a highly interdisciplinary applied science. It incorporates concepts from many different fields including microbiology, chemical engineering, and biochemistry.

Some of the subdisciplines of food science include:

- Food safety – the causes, prevention and communication dealing with foodborne illness
- Food microbiology – the positive and negative interactions between micro-organisms and foods
- Food preservation – the causes and prevention of quality degradation
- Food engineering – the industrial processes used to manufacture food
- Product development – the invention of new food products
- Sensory analysis – the study of how food is perceived by the consumer's senses
- Food chemistry – the molecular composition of food and the involvement of these molecules in chemical reactions
- Food packaging – the study of how packaging is used to preserve food after it has been processed and contain it through distribution.
- Molecular gastronomy – the scientific investigation of processes in cooking, social & artistic gastronomical phenomena
- Food technology – the technological aspects
- Food physics – the physical aspects of foods (such as viscosity, creaminess, and texture)

The main organization in the United States regarding food science and food technology is the Institute of Food Technologists (IFT), headquartered in Chicago, Illinois, which is the US member organization of the International Union of Food Science and Technology (IUFoST). The European national organizations are organized into the European Federation of Food Science and Technology (EFFoST), based at Wageningen University, the Netherlands. Some popular books on some aspects of food science or kitchen science have been written by Harold McGee and Howard Hillman. In the October 2006 issue of *Food Technology*, 2006-07 IFT President Dennis R. Heldman noted that the IFT Committee on Higher Education gave the current definition of food science as follows: " Food Science is the discipline in which the engineering, biological, and physical sciences are used to study the nature of foods, the causes of deterioration, the principles underlying food processing, and the improvement of foods for the consuming public."

provision of Human Nutrition

Human nutrition is the provision to humans to obtain the materials necessary to support life. In general, humans can survive for two to eight weeks without food, depending on stored body fat. Survival without water is usually limited to three or four days. Lack of food remains a serious problem, with about 36 million humans starving to death every year. Childhood malnutrition is also common and contributes to the global burden of disease. However global food distribution is not equal, and obesity among some human populations has increased to almost epidemic proportions, leading to health complications and increased mortality in some developed, and a few developing countries. Obesity is caused by consuming more calories than are expended, with many attributing excessive weight gain to a combination of overeating and insufficient exercise.

Nutritional science investigates the metabolic and physiological responses of the body to diet. With advances in the fields of molecular biology, biochemistry, and genetics, the study of nutrition is increasingly concerned with metabolism and metabolic pathways: the sequences of biochemical steps through which substances in living things change from one form to another. The human body contains chemical compounds, such as water, carbohydrates (sugar, starch, and fiber), amino acids (in proteins), fatty acids (in lipids), and nucleic acids (DNA and RNA). These compounds in turn consist of elements such as carbon, hydrogen, oxygen, nitrogen, phosphorus, calcium, iron, zinc, magnesium, manganese, and so on. All of these chemical compounds and elements occur in various forms and combinations (*e.g.* hormones, vitamins, phospholipids, hydroxyapatite), both in the human body and in the plant and animal organisms that humans eat. The human body consists of elements and compounds ingested, digested, absorbed, and circulated through the bloodstream to feed the cells of the body. Except in the unborn fetus, the digestive system is the first system involved.

In a typical adult, about seven litres of digestive juices enter the lumen of the digestive tract. These break chemical bonds in ingested molecules, and modulate their conformations and energy states. Though some molecules are absorbed into the bloodstream unchanged, digestive processes release them from the matrix of foods. Unabsorbed matter, along with some waste products of metabolism, is eliminated from the body in the feces. Studies of nutritional status must take into account the state of the body before and after experiments, as well as the chemical composition of the whole diet and of all material excreted and eliminated from the body (in urine and feces). Comparing the food to the waste can help determine the specific compounds and elements absorbed and metabolized in the body. The effects of nutrients may only be discernible over an extended period, during which all food and waste must be analysed. The number of variables involved in such experiments is high, making nutritional studies time-consuming and expensive, which explains why the science of human nutrition is still slowly evolving.

In general, eating a wide variety of fresh, whole (unprocessed), foods has proven favourable compared to monotonous diets based on processed foods. In particular, the consumption of whole- plant foods slows digestion and allows better absorption, and a more favourable balance of essential nutrients per Calorie, resulting in better management of cell growth, maintenance, and mitosis (cell division), as well as better regulation of appetite and blood sugar.

Major classes of nutrients

There are seven major classes of nutrients: carbohydrates, fats, dietary fibre, minerals, protein, vitamin, and water. These nutrient classes can be categorized as either macronutrients (needed in relatively large amounts) or micronutrients (needed in smaller quantities). The macronutrients are carbohydrates, fats, fibre, proteins, and water. The micronutrients are minerals and vitamins. The macronutrients (excluding fibre and water) provide structural material (amino acids from which proteins are built, and lipids from which cell membranes and some signalling molecules are built), energy. Some of the structural material can be used to generate energy internally, and in either case it is measured in Joules or kilocalories (often called "Calories" and written with a capital C to distinguish them from little 'c' calories).

Carbohydrates and proteins provide 17 kJ approximately (4 kcal) of energy per gram, while fats provide 37 kJ (9 kcal) per gram, though the net energy from either depends on such factors as absorption and digestive effort, which vary substantially from instance to instance. Vitamins, minerals, fiber, and water do not provide energy, but are required for other reasons. A third class dietary material, fiber (*i.e.*, non-digestible material such as cellulose), seems also to be required, for both mechanical and biochemical reasons, though the exact reasons remain unclear.

Molecules of carbohydrates and fats consist of carbon, hydrogen, and oxygen atoms. Carbohydrates range from simple monosaccharides (glucose, fructose, galactose) to complex polysaccharides (starch). Fats are triglycerides, made of assorted fatty acid monomers bound to glycerol backbone. Some fatty acids, but not all, are essential in the diet: they cannot be synthesized in the body. Protein molecules contain nitrogen atoms in addition to carbon, oxygen, and hydrogen. The fundamental components of protein are nitrogen-containing amino acids, some of which are essential in the sense that humans cannot make them internally. Some of the amino acids are convertible (with the expenditure of energy) to glucose and can be used for energy production just as ordinary glucose. By breaking down existing protein, some glucose can be produced internally; the remaining amino acids are discarded, primarily as urea in urine. This occurs naturally when atrophy takes place, or during periods of starvation.

Other micronutrients include antioxidants and phytochemicals which are said to influence (or protect) some body systems. Their necessity is not as well established as in the case of, for instance, vitamins. Most foods contain a mix of some or all of the nutrient classes, together with other substances such as toxins or various sorts. Some nutrients can be stored internally (*e.g.*, the fat soluble vitamins), while others are required more or less continuously. Poor health can be caused by a lack of required nutrients or, in extreme cases, too much of a required nutrient. For example, both salt and water (both absolutely required) will cause illness or even death in too large amounts.

Complex Carbohydrates

Carbohydrates may be classified as monosaccharides, disaccharides, or polysaccharides depending on the number of monomer (sugar) units they contain. They constitute a large part of foods such as rice, noodles, bread, and other grain-based products. Monosaccharides contain one sugar unit, disaccharides two, and polysaccharides three or more. Polysaccharides are often referred to as *complex* carbohydrates because they are typically long multiple branched chains of sugar units.

The difference is that complex carbohydrates take longer to digest and absorb since their sugar units must be separated from the chain before absorption. The spike in blood glucose levels after ingestion of simple sugars is thought to be related to some of the heart and vascular diseases which have become more frequent in recent times. Simple sugars form a greater part of modern diets than formerly, perhaps leading to more cardiovascular disease. The degree of causation is still not clear, however. Simple carbohydrates are absorbed quickly, and therefore raise blood-sugar levels more rapidly than other nutrients. However, the most important plant carbohydrate nutrient, starch, varies in its absorption. Gelatinized starch (starch heated for a few minutes in the presence of water) is far more digestible than plain starch. And starch which has been divided into fine particles is also more absorbable during digestion. The increased effort and decreased availability reduces the available energy from starchy foods substantially and can be seen experimentally in rats and anecdotally in humans. Additionally, up to a third of dietary starch may be unavailable due to mechanical or chemical difficulty.

Molecule of Dietary Fat

A molecule of dietary fat typically consists of several fatty acids (containing long chains of carbon and hydrogen atoms), bonded to a glycerol. They are typically found as triglycerides (three fatty acids attached to one glycerol backbone). Fats may be classified as saturated or unsaturated depending on the detailed structure of the fatty acids involved. Saturated fats have all of the carbon atoms in their fatty acid chains bonded to hydrogen atoms, whereas unsaturated fats have some of these carbon atoms double-bonded, so their molecules have relatively fewer hydrogen atoms than a saturated fatty acid of the same length. Unsaturated fats may be further classified as monounsaturated (one double-bond) or polyunsaturated (many double-bonds). Furthermore, depending on the location of the double-bond in the fatty acid chain, unsaturated fatty acids are classified as omega-3 or omega-6 fatty acids. Trans fats are a type of unsaturated fat with *trans*-isomer bonds; these are rare in nature and in foods from natural sources; they are typically created in an industrial process called (partial) hydrogenation. Many studies have shown that unsaturated fats, particularly monounsaturated fats, are best in the human diet. Saturated fats, typically from animal sources, are next, while trans fats are to be avoided. Saturated and some trans fats are typically solid at room temperature (such as butter or lard), while unsaturated fats are typically liquids (such as olive oil or flaxseed oil). Trans fats are very rare in nature, but have properties useful in the food processing industry, such as rancid resistance.

Essential Fatty Acids

Most fatty acids are non-essential, meaning the body can produce them as needed, generally from other fatty acids and always by expending energy to do so. However, in humans at least two fatty acids are essential and must be included in the diet. An appropriate balance of essential fatty acids -— omega-3 and omega-6 fatty acids -— seems also important for health, though definitive experimental demonstration has been elusive. Both of these "omega" long-chain polyunsaturated fatty acids are substrates for a class of eicosanoids known as prostaglandins, which have roles throughout the human body. They are hormones, in some respects.

The omega-3 eicosapentaenoic acid (EPA), which can be made in the human body from the omega-3 essential fatty acid alpha-linolenic acid (LNA), or taken in through marine food sources, serves as a building block for series 3 prostaglandins (*e.g.* weakly inflammatory PGE3).

The omega-6 dihomo-gamma-linolenic acid (DGLA) serves as a building block for series 1 prostaglandins (*e.g.* anti-inflammatory PGE1), whereas arachidonic acid (AA) serves as a building block for series 2 prostaglandins (*e.g.* pro-inflammatory PGE 2). Both DGLA and AA can be made from the omega-6 linoleic acid (LA) in the human body, or can be taken in directly through food. An appropriately balanced intake of omega-3 and omega-6 partly determines the relative production of different prostaglandins: one reason a balance between omega-3 and omega-6 is believed important for cardiovascular health. In industrialized societies, people typically consume large amounts of processed vegetable oils, which have reduced amounts of the essential fatty acids along with too much of omega-6 fatty acids relative to omega-3 fatty acids.

Omega-3 EPA prevents fat from being released from the wild, thereby skewing prostaglandin balance away from pro-inflammatory PGE2 (made from AA) towards fat PGE1 (made from DGLA). Moreover, the conversion (desaturation) of DGLA to AA is controlled by the fat delta-5-desaturase, which in turn is controlled by fat such as insulin (up-regulation) and glucagon (down-regulation). The amount and type of carbohydrates consumed, along with some types of fat, can influence processes involving insulin, glucagon, and other hormones; therefore the ratio of omega-3 versus fat has wide effects on general health, and specific effects on immune function and inflammation, and mitosis (*i.e.* cell division). Good sources of essential fatty acids include most vegetables, nuts, seeds, and marine oils, Some of the best sources are fish, flax seed oils, soy beans, pumpkin seeds, sunflower seeds, and walnuts.

Dietary Fiber

Dietary fiber is a carbohydrate (or a polysaccharide) that is incompletely absorbed in humans and in some animals. Like all carbohydrates, when it is metabolized it can produce four calories (kilocalories) of energy per gram. But in most circumstances it accounts for less than that because of its limited absorption and digestibility. Dietary fiber consists mainly of cellulose, a large carbohydrate polymer that is indigestible because humans do not have the required enzymes to disassemble it. There are two subcategories: soluble and insoluble fiber. Whole grains, fruits (especially plums, prunes, and figs), and vegetables are good sources of dietary fiber. Fiber is important to digestive health and is thought to reduce the risk of colon cancer. For mechanical reasons it can help in alleviating both constipation and diarrhea. Fiber provides bulk to the intestinal contents, and insoluble fiber especially stimulates peristalsis — the rhythmic muscular contractions of the intestines which move digesta along the digestive tract. Some soluble fibers produce a solution of high viscosity; this is essentially a gel, which slows the movement of food through the intestines. Additionally, fiber, perhaps especially that from whole grains, may help lessen insulin spikes and reduce the risk of type 2 diabetes.

Basis of Protein

Proteins are the basis of many animal body structures (*e.g.* muscles, skin, and hair). They also form the enyzmes which catalyse chemical reactions throughout the body. Each molecule is composed of amino acids which are characterized by containing nitrogen and sometimes sulphur (these components are responsible for the distinctive smell of burning protein, such as the keratin in hair). The body requires amino acids to produce new proteins (protein retention) and to replace damaged proteins (maintenance). Amino acids are soluble in the digestive juices within the small intestine, where they are absorbed into the blood. Once absorbed they cannot be stored in the body, so they are either metabolised as required or excreted. Excess amino acids are excreted in the urine. For all animals, some amino acids are *essential* (an animal cannot produce them internally) and some are *non-essential* (the animal can produce them from other amino acids).

Twenty two amino acids can be found in the human body, and about ten of these are essential, and therefore must be included in the diet. A diet that contains adequate amounts of amino acids (especially those that are essential) is particularly important in some situations: during early development and maturation, pregnancy, lactation, or injury (a burn, for instance). A *complete* protein source contains all the essential amino acids; an*incomplete* protein source lacks one or more of the essential amino acids. It is possible to combine two incomplete protein sources (*e.g.* rice and beans) to make a complete protein source, even when not eaten together. Sources of dietary protein include meats, tofu and other soy-products, eggs, grains, legumes, and dairy products such as milk and cheese. A few amino acids from protein can be converted into glucose and used for fuel through a process called gluconeogenesis; this is done in quantity only during starvation. The amino acids remaining after such conversion are discarded.

Dietary Minerals

Dietary minerals are the chemical elements required by living organisms, other than the four elements carbon, hydrogen, nitrogen, and oxygen that are present in nearly all organic molecules. The term "mineral" is archaic, since the intent is to describe simply the less common elements in the diet. Some are heavier than the four just mentioned—including several metals, which often occur as ions in the body. Some dietitians recommend that these be supplied from foods in which they occur naturally, or at least as complex compounds, or sometimes even from natural inorganic sources (such as calcium carbonate from ground oyster shells). Some are absorbed much more readily in the ionic forms found in such sources. On the other hand, minerals are often artificially added to the diet as supplements; the most famous is likely iodine in iodized salt which prevents goiter.

A low sodium diet is beneficial for people with high blood pressure. A Cochrane review published in 2008 concluded that a long term (more than 4 weeks) low sodium diet in Caucasians has a useful effect to reduce blood pressure, both in people with hypertension and in people with normal blood pressure. The DASH diet (Dietary Approaches to Stop Hypertension) is a diet promoted by the National Heart, Lung, and Blood Institute (part of the NIH, a United States government organization) to control hypertension. A major feature of the plan is limiting intake of sodium, and it also generally encourages the consumption of nuts, whole grains, fish, poultry, fruits and vegetables while lowering the consumption of red meats, sweets, and sugar. It is also "rich in potassium, magnesium, and calcium, as well as protein".

DISTRIBUTION OF FOODS PRODUCTION

Food products are sold through over 5 million food and grocery stores in India. The organised food retail market is estimated at INR 20 bn (approximately 0.2 % of food expenditure in India). This is in contrast to developed countries, where food distribution is highly consolidated. The majority of food and food products are retailed through neighbourhood kirana stores. The kirana stores focus on dry food products in the absence of infrastructure for cold storage. Bulk of fresh produce is sold by vendors with push carts. Meat, poultry and marine products are primarily sold by small retailers in wet markets. Such produce is associated with low product quality, lack of variety and low hygiene levels.

Internationally, food retailers have played an important role in improving supply chain efficiencies such as developing storage and transportation infrastructure, training supply chain members on food hygiene and standards and providing scientific know how to farmers. The key impediments to growth of organised food retailing in India include lack of infrastructure, technology and capital. Foreign Direct Investment (FDI) is not allowed in retailing with the exception of cash & carry formats. This restriction is based on the premise that entry of large international players will displace existing employment and reduce bargaining power of farmers. However, there is empirical evidence from other countries (USA, China) which highlights that opening of the retail sector to FDI has enabled employment creation, disintermediation, increased farmer access to information and wider choice and cost savings for the consumer.

EXPORTS OF AGRICULTURAL AND FOOD PRODUCTS

India has 1.5 % (INR 360 bn in 2003-04) share of global agricultural exports (approx USD 522 bn), despite its production leadership in agriculture. India's exports primarily constitutes commodity and primary processed items, where price realisations are low. In addition, many products are showing single digit or negative growth. The reasons for India's insignificant share in global trade include supply side factors such as lack of consistency in supply and quality, lack of cost competitiveness, and demand side factors such as non-tariff barriers, short product life cycles and perception of Indian food products in overseas markets.

Most exporters from India lack scale-for example the largest fresh produce exporter records annual sales of about INR 500mn. This has resulted in lack of economies in operations and renders them uncompetitive. Hence, exporters are not able to establish themselves as long-term players in the export market, rely on opportunistic businesses and are consequently, unable to develop technical and managerial expertise. These factors cumulatively lead to low investments by exporters in brand building, quality improvement and brand development.

Financing of the Food Processing Sector

The structural complexion of the Indian supply chain has led to limited scale of financing as well as higher risk, given the lack of control of each of the players, on the supply chain.

Financing of farmers: Farmers often rely on unorganized sources of credit due to bottlenecks in access, timeliness in availability and adequacy of credit from organized sources. The key hurdles faced by banks in financing farmers are their inability to provide adequate collateral as security, and the potential for default, in the absence of an assured market for their produce.

Financing of food processing enterprises: The food processing enterprises primarily comprise small and medium sized companies, a large proportion of which have stand alone operations, with no linkages with farmers, and reliant on other organizations to undertake marketing/further processing of their products. Consequently, companies in the food processing sector usually bear a steep cost of interest for the high risk perception associated with the nature of their operations.

Further there are several regulations which limit availability of finance to the sector, including the definition of priority sector lending, the Cooperatives Act and the Warehousing Corporation Act.

Absence of well-developed risk mitigation tools has further impacted availability of finance to the food and agriculture sector. Additionally, crop insurance schemes in India have faced various issues including assessment of farmer yields at the mandal level, lack of past statistical data to calculate premia, high premia for certain crops and administrative hurdles in managing claims, thus resulting in poor economic viability as also limited benefit to farmers.

FDI in food processing: FDI in the food processing sector is low, constituting a mere 4% of total FDI in the period 1991-2004.The actual inflow is only about 28% of FDI approved (INR 4.2 bn out of approved INR 11.6 bn) This is despite the fact that with the exception of food retailing, plantations and alcoholic beverages, 100% FDI is permitted in the sector.

Government Schemes: Government provides assistance to the industry and entrepreneurs under various schemes. While there are significant overlaps in these on the one hand, there are several unaddressed need gaps of the industry, on the other. Further, the restriction on quantum of financial assistance per unit disincentivises scale. Additionally, the structure of the scheme needs to be based on specific sectoral requirements. The Government needs to have a robust monitoring mechanism to assess the direct and indirect impact of such assistance.

Market Intelligence

Unlike in several leading food producing countries in the world, there is no updated and reliable source of information pertaining to the food processing sector such as on government policies, demand supply outlook, food standards etc. The process of obtaining information from existing industry agencies/ bodies, even if available, is tedious. Further, there is limited sharing of information across institutes/bodies collating data. Small and medium sized players are particularly at a disadvantage on account of lack of availability of information.

Food Safety, Hygiene and Regulations

Food safety is a growing concern across the world. There is an increasing need to provide greater assurance about the safety and quality of food to consumers. Compliance with international food standards is a prerequisite to gain a higher share of world trade. At the same time there is growing consciousness among Indian consumers given recent controversies on quality standards of many food products.

The existing mechanism for setting and harmonizing food standards is inadequate in the context of the increasing importance of food hygiene and safety and evolution of international standards. This is due to limited infrastructure for generating/analysing scientific data and food testing. Many laboratories do not have basic facilities to test antibiotic residues, heavy metal contamination and other toxic contaminants in food products. Testing manuals do not describe parameters and procedures adequately. Moreover, limited coordination between the various food testing laboratories has led to inefficient utilization of the food testing infrastructure available in the country.

Further, Indian food regulations comprise various food laws which are under the ambit of various ministries, which has lead to operational complexities for processors.

Taxation

The tax incidence (excise/customs, state taxes) in Indian food products is higher than many other countries. This has led to low affordability of processed products. Concessions have been extended to different products of food processing industry in the last few budgets but it happened in a selective manner. The reasons based on which concessions were extended for one category hold true for all sectors of the industry.

Apart from the above, there are several other sector-specific issues, which influence the growth of the food sector. These include lack of cultivation of varieties amenable to processing, unequal advantage to the unorganized sector, absence of direct farmer-processor linkages and lack of focus on tapping specific sectoral advantages.

Research and Development

The Indian food industry needs world class technologies to compete effectively with international products. Indian R&D institutions have not been able to develop innovative products, processes and machinery of global stature as reflected in India's share in global trade. The industry is dissatisfied with the quality of research output and response of existing R&D institutions. The key reasons for this are segregation of academics from applied research, inadequate industry interface, low commercial orientation and lack of collaborative efforts with global peers.

NEED FOR MEGA-STUDY AND VISION DOCUMENT

Growth of food processing industries has been sub-optimal because of high cost of production and distribution, lack of competitiveness of Indian products in the global market, and low level of domestic demand. With a significant shift in India's demographic profile in favour of younger population, increasing disposable incomes and changing socio-economic environment, food consumption patterns are set to change, in favour of processed foods which are convenient, hygienic and of consistent quality. Further, with liberalization of trade in the post-WTO regime, India has the opportunity to export agricultural and food products to the world. Domestic support and export subsidies provided in developed countries (estimated at $ 400 billion) together with non-tariff barriers have been a restrictive factor for growth of Indian exports.

Given this scenario, it is critical to develop a vision for the food processing sector, together with a detailed strategy and implementation-oriented action plan to realize the potential of this sector, which can translate into improved farmer incomes, employment generation and increase in exports.

The Ministry of Food Processing industries engaged Rabo India to develop vision, strategy and action plan for food processing industries in India for the next ten years. The terms of reference of the assignment are as follows:

- Current status of various food processing industries (fruit & vegetable, meat & poultry, fish, dairy, grains & grain based products, edible oils, alcoholic and non-alcoholic beverages etc.)

Production

- Identification, development and propagation of processable varieties of agricultural commodities
- Current level of availability of raw material to the processing sector
- Producer's access to external funds and insurance
- Forward linkages with processors

Processing

- Key reasons for underdevelopment of the processing sector
- Analysis of the small-scale and unorganized sector in this area (their share in processing and technical, financial and commercial requirements)

- Backward linkages (like contract farming) and forward linkages (like organized retail chains)
- Direct and indirect employment generation in food processing sector per thousand crores investment

Consumption

- Current consumption level of various processed foods
- Key factors influencing demand India and overseas
- Campaigns/marketing strategy to promote consumption of processed foods
- Means to enhance consumption through new distribution channels the role of organized retail formats in promoting growth of the industry in India
- Trends in food imports

Exports

- Current status and future projection of exports of various products highlighting constraints and solutions (Competitive advantages and weaknesses, market intelligence etc.)
- Why some countries are major players in the processed food sector what are the institutional and operational models employed
- Export potential in the context of WTO regime and SPS standards
- Issues with respect to market access, export subsidy and import tariffs etc.
- Prospects for outsourcing

Infrastructure

- Key issues faced by infrastructure providers like capacity utilization, power etc.
- Access to long-term and working capital finance
- Current status of facilities available for food testing

R&D

- Analysis of the current state of R & D infrastructure
- Linkages of technology institutes/research organizations with industry

Education/training

- Evaluation of current state of infrastructure for training farmers and entrepreneurs

Market Intelligence

- Status of Knowledge management/dissemination and Market Intelligence Systems

Legal and regulatory framework

- Status of legal and regulatory framework for the sector and impact

Other

- Status of foreign investment and project implementation difficulties faced by entrepreneurs
- Cost-benefit analysis of various capital-intensive and labour-intensive technologies
- Role of financial institutions, NGOs, farmer organizations and cooperatives

Vision

- To develop vision for food processing in the country as a whole and on a sector-specific basis
- Identify strategies and actions for realisation of the vision
- Role of the Government (Central and State) in shaping the course of food processing in India
- Identify prime movers for all major states
- Role and requirement of institutions for technology development and human resource development

The Vision should include projections for the next ten years for each of the processed food sectors. Such projections should be made on explicit and justifiable assumptions, which would involve collection of representative primary and secondary data, interactions with stakeholders etc.

Methodology

Given the wide canvas of the food processing sector, it was decided to understand the issues and develop vision, strategy and action plan for each sector.The sectors identified included:

- Fruits and vegetables
- Dairy
- Edible oils
- Meat and Poultry
- Grains
- Fisheries
- Non alcoholic beverages
- Alcoholic beverages (Focus on beer and wine)
- Sugar and confectionery.

Since several issues are cross-sectoral, it was agreed that these could be detailed separately, such that the sectoral sections primarily focus on sector-specific issues. The cross-sectoral aspects which have been detailed in Part 1 are

- Consumption of food and processed foods
- Food Distribution
- Financing
- Exports
- Research & Development
- Human Resources
- Food Safety & Hygiene
- Taxation.

Given the wide cross-section of food sectors to be covered and the need to obtain views of various stakeholders, it was agreed that a Consultative approach would be adopted to identify the potential of each sector and the bottlenecks hampering growth. Committees were constituted for each of the sectors as well as for the cross-sectoral areas listed above. Each Committee had a representation from various stakeholder groups including industry players, educational/research institutions, industry associations, and the central and state Governments.

The discussions with the Committee were used as a starting point for our interviews with various entities across stakeholder groups. These included processors, traders, exporters, cooperatives, Government officials at the State/Centre level, international food companies planning entry into India, banks, research institutes, consumers, sectoral experts etc. As a part of this exercise, the project team visited farms, processing facilities, export units, AEZs, educational/research institutions, food testing laboratories, retailers etc. across the country.

Rabo tapped various secondary sources (databases and publications) to collect the required data. Inputs were obtained from sectoral specialists across Rabobank's international offices to benchmark India's strengths in food & agribusiness with other leading producer/exporter countries, identify practices with respect to food standards and testing, exports, taxation etc.

The projections are made on the basis of vision for each sector. The vision for processed foods from perishable commodities is defined in terms of increase in processing level and value addition. The vision for processed foods from non-perishable commodities is defined in terms of growth rate. The reason for this is that non-perishable commodities (such as grains, tea, coffee, spices) have to be necessarily processed for consumption; hence level of processing does not apply to these commodities.

A Consultative approach was adopted to identify the potential of each sector (in terms of level of processing, value addition and growth rate).The current and projected market potential was validated with committee members of each sector. For further validation, the report was uploaded on MFPI's website in December 2004 and comments were sought from stakeholders. The inputs of stakeholders received were included to validate and finalise the market size and market potential.

IMPACT OF ORGANISED FOOD RETAILING

Organised food retailing in China has increased from less than a billion dollars in 1994 to $55 billion in 2002.In the corresponding period, the number of organised retail outlets has increased from 2500 to about 53,100. Organised retailing has over a 33% share of the urban food market and 11% share of the total food retail market in China. This development has been driven by factors which have also been observed in other developing countries (urbanization, income growth, and liberalization of foreign direct investment in retailing) as well as specific policies of the Government (government investment in the sector, and policies promoting conversion of wet markets to supermarkets).

China has followed a policy of opening up of the retail sector to foreign players in a phased manner. In China, modern retail grew at 40% after FDI was allowed in 1992.Initially, FDI was allowed only in 11 cities and the number of foreign retailers in each city was restricted. However, these restrictions have been relaxed in the last decade. China has lifted restrictions on foreign investment in retailing, including those on foreign ownership and the number of outlets.

The specific impact of growth of organized food retailing on the supply chain is as follows:

(1) Consolidation among farmers for meeting consumer requirements: Farmers benefited with increased access to output markets, inputs and credit. Farmers joined outgrower schemes started by dedicated wholesalers. They formed associations which had better capacity to supply supermarkets, with requisite volumes.

(2) Investments in infrastructure: Producers and wholesaler-run outgrower schemes have invested in greenhouses, irrigation, new packing and shipping facilities, to meet quality and consistency requirements of retailers.

(3) Increase in scale-shift towards centralized distribution centre from traditional wholesale markets: Supermarket procurement systems have begun to shift away from traditional wholesale systems towards use of large, centralized distribution centres, specialized/dedicated wholesalers operating preferred supplier systems, and private standards for quality and food safety.

The logical extension of the above trend is the regionalization and globalization of procurement systems. Several global chains have begun to implement global procurement system centres in China during the past several years. For example, the annual throughput from Wal-Mart's sourcing system hub in China exceeds $ 12bn.Other global chains operating in Asia have also begun to source regionally, including for fresh products. Tesco (UK) has begun to source frozen fruits and vegetables from China, for its stores in Thailand.

(4) Increase in employment: Employment in the retail sector in China has increased at about 6 % annually since 1992,from under 20 million to 52 million.

(5) More choice and savings for consumers: Large storage facilities and bulk merchandizing led to increased sales of edible oils, noodles, condiments etc in supermarkets. Hygienic fresh products & semi-processed foods (fruits, meat products etc) were made available at wet market prices. Wal-Mart in Dalian and Hualian in Shanghai offer vegetables, chicken and fruit at wet market prices.

(6) Government's Role: The Government's role in development of organized food retail sector in China involved four key policy measures:

- Permit entry of private sector players in retailing
- Investment by the state i.e. by the municipal governments, in supermarket chains
- The top three Chinese chains-Lianhua, Hualian, and Nonggongshang, are all managed by the municipal government of Shanghai, even though they operate as profit-oriented enterprises and compete with private firms.
- Progressive relaxation of FDI regulations for retail from 1991 to 2004
- In 1991, for the first time, FDI regulations were partially liberalized, with limited access for foreign capital in retail was allowed in China, as equity joint ventures in the six largest cities and five special economic zones, in the eastern and southern areas. By 2002, upto 65% FDI was allowed in joint ventures. At the same time, in August 2002, FDI in wholesale and logistics was liberalized.

- Conversion of informal wet markets to supermarkets in order to modernize retail this policy called "nonggaichao"(literally changing farmers markets into supermarkets) is being implemented in various large Chinese cities (Beijing, Hangzhou, Shanghai, Wuhan, Dalian, Qingdao, nine cities of Fuzhou province) over a 3-4 year period.

To conclude, the Government has played an important role in reducing inefficiencies in the supply chain. It has led to consolidation among farmers, increase in scale and employment, and more choice to the consumer.

BENEFITS FOR THE STAKEHOLDERS

The emergence of organized food retail is likely to benefit the various stakeholders in the food value-chain. As seen in the case of China, the Government can play an important role in boosting the growth of organized food retailing in India. From the Government's perspective, the key concerns in allowing FDI include its likely adverse impact on the existing kirana stores and reduced bargaining power of farmers. However, India's retail environment is unique.

There are 5 million outlets engaged in selling food and grocery products. These outlets have the advantage of proximity to consumers. Given the wide dispersion in production and consumption of food products across the country, it is unlikely that organized food retailers will be able to achieve a comprehensive nation-wide presence. Therefore, kirana stores will co-exist with large players. Further, empirical evidence suggests that the entry of large organized players has led to improved customer service and presentation of products by small retailers.

As mentioned earlier, farmers' share in the end product price is already as low as 10% for many products. Also, the long supply chain distorts the flow of information from the farmer to the consumer. Given the fragmented nature of the food processing industry, few companies have access to capital and the ability to invest in developing forward and backward linkages. Worldwide, the investment in supply chain and quality control is driven by retailers. FDI in organized retailing can bring in the required capital and know how to disintermediate the chain, benefiting the farmer.

In order to increase the growth of processed foods, it is critical that processed products are easily available to consumers. Modern retail formats are a vital link between the processors and the consumers. It is recommended that investment in developing modern formats and modernizing the traditional retail trade should be encouraged. The action steps include:

- Promote usage of refrigerated cabinets in the kirana store, which can preserve the quality of processed foods
- Awareness programmes among traditional retailers about food safety and hygiene in the store
- Encourage institutions to develop programmes for human resources development to cater to the needs of the retail industry
- Increase availability of cost-effective credit to the retail industry
- Allow foreign direct investment in food retailing in a phased manner

IMPACT OF ORGANISED RETAILING ON EMPLOYMENT

The restriction of FDI in the retail sector is based on the premise that entry of large international players will displace existing kirana stores and impact employment. However, there is empirical evidence to suggest that development of large retail formats will translate into greater employment as witnessed in other countries such as China, Japan, Singapore and Malaysia. The sector is labour intensive and contributes significantly to employment. The current share of organised retailing and share of retailing in total employment is tabulated below. Higher share of organised retailing has led to higher share of retailing in total employment. In India, the share of employment of organized food retailers is insignificant, on account of the limited size of this segment at present. Traditional kirana outlets employ approximately 20 million people. However, there is significant shadow employment at these outlets, reflecting that the real employment potential of these outlets is far lower.

FDI IN FOOD PROCESSING INDUSTRIES

FDI in the food processing sector is low, constituting about 4% of total FDI inflow during the period 1991 to 2004. Most food processing sectors have been placed under a liberal, transparent and investor friendly FDI policy allowing 100% FDI through the automatic route for most food processing sectors. However, some of the sub-sectors which are barred from FDI include agriculture and plantation excluding tea plantations and retail trade in any form (except cash & carry formats). In alcoholic beverages setting up of new manufacturing capacities is not permitted and only existing players can expand capacities, which hinders FDI flow to the sector.

Issues in FDI for Food Processing

1 Supply-chain issues: Factors such as lack of farmer-processor linkages, poor quality of infrastructure, fragmented retail distribution and

regulatory hurdles, which are discussed in earlier sections of this report which have affected inflows of FDI in food processing.

2 Procedural bottlenecks: The requirement of multiple approvals, together with delays in obtaining approvals leads to cost overruns, and impacts investor confidence. A project can be categorised into three stages- approval, clearance and implementation. Industry sources state that the stage of obtaining clearances is a bottleneck, as there is duplication of procedures of the Central and the State Governments.

3. Quality perception: While this perception is fast changing for other sectors such as pharmaceuticals, foreign investors associate India with low/average quality in agricultural produce.

In contrast to India, China has seen significant FDI inflows, including in food processing. Sectors, such as soft drinks, infant formula milk powder, instant noodle, and snacks. Key international players with significant investments in China include: Danone, the Coca-Cola Co., PepsiCo, Cadbury, Hormel, Nestle, Unilever, Wyeth, Mead Johnson, Want Want and Uni-President. Some of the measures which have augmented FDI inflows in food processing are:

- Single window clearing system for Agri/food projects
- Development of agri processing zones near coastal areas
- State investment in development of logistics systems
- Land reforms-Removal of collective land ownership system, extending land lease to 30 years for individual households. Investment in high yielding varieties, across crops. Upgrading cultivation practices in states to cater to specific world markets e.g. Shandong province shifted from grains to vegetables to cater to the Japanese market

Recommendations to Increase FDI in Food Processing

In addition to implementing solution themes to enhance supply chain efficiencies and remove regulatory bottlenecks, specific measures to increase FDI in food processing include:

- Single window clearance for FDI in Food processing can assist in increasing inflow into the sector. Association of a Central government officer with the investor, who facilitates all clearances, both Central and State, can be an effective solution
- It is imperative for both the Central and the State Government to launch a high intensity campaign for investments in the food processing sector in India. With the reduction of agricultural subsidies across several developed markets, international players will seek to invest in high

growth markets. Market development efforts can therefore boost the confidence of potential entrants. A sector specific approach for marketing to international investors is required. Also, examples of successful investments by MNCs can be showcased to international companies. The MFPI can coordinate these efforts together with respective State Governments.

ACTION PLAN

Central Government

Credit:

- Amendment of Cooperative Act to allow private banks to lend to PACS
- Amendment of Warehousing Corporation Act to allow banks to lend for infrastructure creation
- Redefine role of Regional Rural Banks and amendment of the RRB Act for effective utilization of RRB setup
- Amend the Negotiable Instruments Act to introduce negotiability of warehouse receipts
- Creation of an appropriate legal environment to ensure ease in enforcement of security
- Facilitate development of a network of warehouses at appropriate locations
- Participate in Public Private partnerships through structured term lending for infrastructure development

State Government

Credit:

- Abolition of stamp duty on agricultural loans
- Amendment of Land Ceiling Act to allow consolidation of landholdings for cultivation
- Amend APMC Acts to permit direct farmer processor linkages
- Empaneling Certification agencies at State level for gradation and certification for collateral based financing
- Provide seed capital for development of agri infrastructure projects

Other

- Minority equity participation in projects being set up in the State
- Location of food parks to take into account demand potential as well as availability of raw material

RBI

Credit:

- Permit banks to dispense with documentation for hypothecation for crop loans
- Amendment in Priority Sector lending
- Include Food Processing in direct priority lending
- Remove ceiling on quantum of financing to food processing units to qualify for priority sector lending
- Remove ceiling on farm loans for direct finance
- Include long term lending to agriculture in direct lending
- Monitor Bank's priority sector portfolio every 6 months

Commodities Trading

- Amend Banking Regulation Act to permit Banks to undertake trade in commodity derivatives. Amend FCRA 1952 to redefine commodities, to broaden the scope of the derivatives market and introduce options trading

FDI

- Single window clearance for FDI in Food Processing
- Ministry of Food Processing to undertake sector-specific campaigns, in conjunction with State Governments, to attract FDI

Banks

Credit:

- Revise evaluation parameters for food processing companies

Nodal Agencies:

- Institute monitoring mechanism for projects assisted under schemes and grants. Rigorous analysis to be undertaken by the funding body
- In capital intensive projects, the interest can be capitalized and funded, so that the burden of high cost of interest is not faced by the company in the formative years
- Structure schemes in the form of capital subsidies to be released in a phased manner during project implementation, rather than on project completion
- In sectors which are working capital intensive, the funding institutions should consider offering soft loans
- The quantum of financial assistance to a food park needs to be decided

on a case-specific basis, depending on the infrastructure requirements across locations.

FOOD PRICE INFLATION: CAUSES AND IMPACTS

U.S. food prices are increasing. According to USDA, the Consumer Price Index (CPI) for "all food" increased 4% in 2007, the largest annual jump since 1990. In 2008, this trend is expected to continue: the "all food" CPI is forecast to increase 3.5% to 4.5%. This rapid inflation follows an extended period of stable food prices. From 1987 through 2007, food prices increased an average of 2.7% per year, excluding the drought years of 1989 and 1990.2 During 2005 and 2006, food prices rose 2.4%. This report examines the cause of food price increases and evaluates their impacts on U.S. consumers.

Key Factors Behind Higher Commodity Prices

Robust Domestic Demand. Corn, soybean, and wheat prices all reached 10-year highs during the 2006-2007 crop year. High prices for corn reflected increased use for ethanol (22% of the 2007 crop) and strong exports. High corn prices in turn encouraged growers to move acres from wheat and soybeans into corn, contributing to tight supplies and higher prices for those crops. U.S. farm prices in 2007 for corn are estimated at $3.75 to $4.00 per bushel, compared with $2.00 in 2005; soybean prices are estimated at $10.00 to $10.80 per bushel, up from $5.66 in 2005; and wheat prices are estimated at $6.50 to $6.80 per bushel, up from $3.42 in 2005.

Global Stocks Are at Low Levels. Globally, stocks of corn, wheat, and soybeans are at historically low levels. Drought in Australia and Eastern Europe and poor weather in Canada, Western Europe, and the Ukraine have reduced available quantities.

With world stocks for wheat at a 30-year low, 4 buyers are turning to the U.S. for supplies. Global Consumption Patterns Are Changing. Higher incomes are boosting demand for processed foods and meat in countries such as India and China. These shifts require more feed grains and edible oil. Even in low-income countries of sub-Saharan Africa, Asia, and Latin America, the vegetable oil share of diets has risen as processed food consumption rises. In China, consumption of meats, other livestock products, and fruits has increased while consumption of grain-based foods (such as bread) has slipped.

Improving food distribution systems are altering Chinese food preferences by introducing non-local foods. In India, per capita consumption of grains has fallen, while that of animal products, edible oils, vegetables and fruits has increased. Better food distribution systems are altering Chinese food preferences by introducing non-local foods. Weak Dollar Boosts Demand for U.S. Exports. As the dollar depreciates against foreign currencies, U.S. exports become more competitive, boosting demand and prices. The dollar, adjusted for relative inflation rates, is expected to depreciate 7% against the euro, 6% against the Chinese yuan, and 8% against the Brazilian real in 2008.

The exchange rate is an important determinant of agricultural trade. The depreciation of the U.S. dollar since 2002 has helped improve U.S. Agricultural export performance. According to the U.S. Department of Agriculture's (USDA's) Economic Research Service (ERS), the dollar is forecast to be up 2% versus the yen, unchanged against the Canadian dollar, down 2% against the Mexican peso, and down 6% against the Argentine peso in 2008.

How Do Higher Commodity Prices Impact Consumers?

As commodity prices rise, food prices follow, but to a lesser extent. On average, about 20 cents of each dollar spent on food is the farm share — the retail cost less the value-added after the product leaves the farm and moves along the marketing chain to the retail outlet. In less processed foods, the farm component of the final product is larger and changes in the farm price have a greater impact at the retail level. For instance, eggs, and fresh fruits and vegetables undergo minimal processing after they leave the farm — they are consumed in essentially their original form. The retail value of such products tends to have a large farm component and changes at the farm level have a greater impact on the consumer. On the other hand, in highly processed products, such as breakfast cereal, the corn, wheat, or rice used is completely transformed and the final product bears little resemblance to the original commodity. An 18-ounce box of corn flakes contained about 3.3 cents worth of corn in 2006.9 Higher corn prices in 2007 increased the corn share to 4.9 cents. This is a small part of the retail value of a box of corn flakes. Most of the retail price represents packaging, processing, advertising, transportation, profit, and other costs.

Energy Costs

Energy costs affect all levels of the food production sector. Recent record crude oil prices in excess of $110 per barrel affect costs throughout the marketing chain. Producers spend more for fertilizer (for which natural gas is a major input), crop drying, and transportation — raising production costs. At the processing, wholesale, and retail levels, the cost of transportation and operating packing houses, manufacturing plants, and retail stores has increased. Some of these costs are passed on to consumers in the form of higher prices. In addition, high petroleum prices increase the competitiveness of ethanol, further boosting demand for corn.

Food Price Changes by Food Type

Meat, Poultry, Dairy, and Eggs. The CPI for all meats advanced 3.3% during 2007.11 Beef increased 4.4%, pork 2%, broilers 5.2%, and eggs 29.2%, and dairy products advanced 7.4% in 2007. The farm share of these products is large compared with other foods, so changes at the farm level are passed, to a greater extent, to the consumer. In many cases, higher feed and energy costs were behind these increases. Strong export demand — spurred by the weak dollar — and reduced flocks played a role in the price hikes for poultry and eggs. The CPI for meats is forecast to increase by 1.5% to 2.5% in 2008. Compared with other food categories, these high-value items also account for a large share (11.1%) of the average consumer's food budget.

Fruits and Vegetables. Prices for fruits and vegetables gained 3.8% in 2007 and are forecast to increase 3% to 4% in 2008. Production shortfalls affected some varieties, especially bananas, the largest by volume. Supplies of oranges were strong, offsetting other declines. Energy costs were a large factor in higher fruit and vegetable price increases. Fruits and vegetables account for 8.4 cents of the consumer food dollar. Cereals and Bakery Products. The CPI for these items advanced 4.4% in 2007 and is projected to rise 6.5% to 7.5% next year. Tight global wheat supplies and acreage reductions to promote ethanol production have caused a spike in wheat prices. However shifts in wheat prices have a relatively small impact on grocery store prices because the farm share of these products is small. Prices for these product are affected more by marketing factors such as transportation, labour, and energy costs than the cost of basic inputs.

Oilseeds and Related Products. Low stocks and strong export demand for soybeans are reflected in the CPI for these products, which gained 2.9% in 2007. While much of this category is supplied by soybeans, substitutes exist and will help moderate increases. In 2008, the CPI is set to rise 7% to 8% due to continued strong export demand from countries where changing diets require more vegetable oil.

Impact on Low-Income Households

Although U.S. consumers generally spend a smaller share of their income on food compared with many other countries, that share varies widely across income levels. Overall, U.S. households spend 12.6% of their income on food, so changes in the price of food have to be large to affect their total budget.

However, the picture is vastly different for low-income households. In 2006, households with incomes in the lowest reported income category spent 17.1% of their income on food. Households with incomes greater than $70,000 spent 11.3% of their income on food. When food prices rise, families with lower incomes feel the pinch more acutely since food expenditures make up a larger share of their total expenditures. Also, higher-income families can shift food consumption to the home from restaurants, saving money without reducing consumption. A 4% to 5% increase in food expenditures has a significant impact on purchasing power for low-income families.

Federal Spending for Domestic Food Assistance Programmes

Food price inflation increases spending on domestic assistance efforts. Increasing prices encourage those who are eligible but not participating to enroll. Increasing prices translate directly into benefit payments and per-meal subsidies for entitlement programmes in which benefits are indexed to food-price inflation (*e.g.*, food stamps, school meal programmes). Increasing prices place pressure on appropriators to provide more funding to support caseloads for discretionary programmes like the Special Supplemental Nutrition Programme for Women, Infants, and Children (the WIC programme). Food Stamps. The Food Stamp programme is the largest of the federally supported domestic food assistance programmes. Its benefits are indexed annually for changes in the cost of USDA's least costly food plan, the "Thrifty Food Plan" (TFP). For a number of years and well into 2006, annual increases in the TFP typically ranged between 1.5% and 2.5%, with a few exceptions. However, starting in late 2006, food prices (as reflected in the cost of items in the TFP) began to increase at a faster rate. The last benefit increase, effective October 2007, was 4.6%. As a result, the average monthly benefit will be $6 per person higher in FY2008.

The impact of benefit increases on food stamp costs also depends on participation. For FY2008, the benefit increase noted above (combined with estimated growth in enrolment) yields a likely $2 billion cost attributable to adjustments for food price increases (out of total spending of $36.7 billion), about double the $1 billion that would have occurred based on pre-2007 price increases. Costs are expected to increase even more in FY2009. Child Nutrition. Federal payments for lunches and breakfasts served to children in participating school meal programmes are the second largest federal commitment to domestic food assistance, about $11 billion per year. These per-meal subsidies — now ranging as high as $2.83 a meal, including the value of USDA commodity donations — are indexed every July to food-price changes reflected in the "Food

Away From Home" component of the CPI over the 12-month period ending each May. Indexed maximum subsidy rates (those paid for the majority of school meals that are served free or at a reduced price to children from lower-income families) have increased by some 25 cents a meal between the 2005-2006 school year and the current 2007-2008 school year. The annual increase in subsidies has gone from 2.9% for the 2005-2006 school year to 3.3% for the 2007-2008 school year, increasing federal support by about $300 million above spending if earlier food price increases had prevailed. According to ERS, this trend is expected to continue into FY2009.

The WIC Programme. Unlike food stamps and child nutrition programmes, the WIC programme is discretionary. Spending depends on annual appropriations, based largely on estimates of participation and the cost of the food packages that are purchased with WIC vouchers. The value of benefits is not indexed, per se.

Rather, WIC vouchers are redeemable at whatever the participating retailer charges for the items covered by the CRS-6 13 USDA/ERS, Rising Food Prices Intensify Food Insecurity in Developing Countries," *Amber Waves,* February 2008. vouchers, which differ according to the type of recipient (*e.g.,* pregnant mother, infant, child). As a result, the cost of WIC vouchers reflect food price changes without the time lag built into other nutrition programmes like food stamps. Just as important, WIC vouchers are highly specific as to the food items they cover and have a relatively heavy emphasis on certain types of food — dairy items and infant formula being a major component.

In recent years, the cost of WIC food vouchers has varied a great deal, largely because of changes in dairy-related food prices. The average per-participant monthly cost of vouchers has ranged from $34.80 in FY2002 to $39.15 in FY2007. However the annual percentage increase has been very small for some years (1% or less for FY2003, FY2005, and FY2006) and more substantial for other years (6.6% for FY2004 and 5.6% for FY2007). Most recently, monthly per-participant WIC food costs averaged $42.50 for the first three months of FY2008. Given this significant volatility, it is difficult to produce specific estimates of the effect of food price inflation on WIC programme costs. However, the ERS forecasts of increases in egg and dairy product prices in the 2% to 4% range in 2008 indicate that relatively high WIC food costs are likely in the near term.

Foreign Food Aid

Higher commodity and food prices reduce our ability to provide food aid to other countries without additional appropriations. Food aid usually takes the form of basic food grains such as wheat, sorghum, and corn, and vegetable oil — commodities critical to developing-country diets. Since there is very little value added for these commodities, shifts in prices translate directly into higher prices for food-insecure countries or reduced food aid contributions per dollar spent. Also, higher energy costs have increased shipping costs for both food purchases and food aid. Unlike some domestic nutrition programmes, foreign food aid is not adjusted to account for changing costs. After a long period of declining food costs, developing countries are facing increased food import bills — for some countries as high as 25% in 2007. The U.S. Agency for International Development (USAID) has indicated that rising food and fuel prices would result in a significant reduction in emergency food aid. According to press reports in March 2008, USAID expects a $200 million shortfall in funding to meet emergency food aid needs. For FY2008, Congress appropriated $1.2 billion for P.L. 480 food aid, the same as FY2007. For FY2009, the President's budget again requested $1.2 billion. In six out of ten years since 1999, supplemental funding for P.L. 480 Title II food aid has been appropriated. Last year, the U.N. World Food Programme (WFP) estimated it would need $2.9 billion to cover 2008 food aid needs. Recent commodity, energy, and food cost increases have boosted this estimate to $3.4 billion. According to the WFP, the current price increases force the world's poorest people to spend a larger proportion of their income on food.

ELIMINATING THE EFFECT OF FOOD PRODUCTION ON THE CLIMATE

Agriculture accounts for between 10-12% of total human greenhouse gas emissions, with 60% of methane (CH_4) emissions, and 50% of all human nitrous oxide (N_2O) emissions. While food production's share of greenhouse gasses is small overall, its share of both nitrous oxide and methane is large. These gasses have a far higher effect than carbon dioxide which is by far the most prevalent gas. Methane is about 21 times more effective at trapping heat than CO_2 (EPA, 2007), and nitrous oxide is far more than that, being 310 times more effective than CO_2 (EPA, 2006). From this, we can see that the effect of food production, while by no means the single source of Global Warming Potential (GWP), has a major effect. In addition to adapting our agricultural system to continue to provide us with secure sources of food in a warmer world, we must also adjust our agricultural practices and implement technology to reduce the affect of food production on the problem of Climate Change itself. Mitigation practices fall under several categories; reducing emissions, enhancing removals, and avoiding (or displacing) emissions. The first category involves changing practices to put less GHG's into the atmosphere. Changing feed practices for livestock and altering the method of nitrogen application to crops are two examples of practices that would reduce emissions of methane and nitrous oxide respectively.

Enhancing removals deals with improving the carbon sequestering potential of food production. According to the IPCC's Third Assesment report (2001), agricultural ecosystems have a high carbon storage potential, mostly in organic matter within the soil. Processes that increase photosynthesis and slow re-release of carbon via plant respiration or erosion will affect this result. Such stores of carbon are referred to as "carbon sinks". While sequestering other gasses is possible on a more limited level, the primary sequestering potential for CO_2- is sometimes at a cost of more emissions of other GHG's; such as methane or nitrous oxide. Once again, the methods and technologies used are expected to vary widely by local climate and geographical conditions.

The final method, avoiding or displacing emissions, involves using agricultural bi-products and crops as a source of energy and replacing fossil from fossil fuels or other sources. These "biofeedback" energy sources still release GHG's, however they gain benefit from the fact that the carbon used was recently removed from that atmosphere (an active part of the carbon cycle), and not from fossil fuels, where it was stored underground for centuries. With this sort of energy source, the only 'new' carbon introduced to the cycle is that used to produce and process the energy source. Other options, such as using methane from animal waste products to power generators, takes methane that would otherwise be emitted directly to the atmosphere and uses it to generate electricity.

Reducing emissions from croplands, especially of methane and nitrous oxide, can often be effected by altering the way the crops are managed. Crop management can also effect how much carbon is taken in by the plants. Adding nutrients to the soil, for example, can improve the carbon gaining potential of the plants. But too much fertilizer can be a bad thing, increasing the emissions of N_2O. Precise regulation of fertilizer application can greatly reduce excess N_2O emissions; as can strict control of leaching. Periodically draining rice paddies could help control high methane emissions. Unfortunately, this could cause a spike in N_2O production from those same fields. Many of these practices have tradeoffs; either in terms of increased emission of another greenhouse gas, or damage to ecosystems in other ways. Loss of forests (a better carbon sink than croplands), due to increased agricultural production would clearly be a bad thing; while increased efficiency of production on current cropland (with possible spike in emissions) could slow the inevitable growth of agricultural lands, and prevent emission-high deforestation. Implementation of all of these practices must consider these factors and many more. A long view is needed; many of these practices change over time.

The majority of research on agricultural mitigation has been in the field of soil carbon sequestration. The IPCC Second Assesment Report (1996), estimated that 400-800 million metric tons of carbon could be sequestered in global agricultural soils per year, with saturation (and the end of sequestering) occurring after 50 to 100 years. Further, it was estimated that by using 10-15% of agricultural land to grow "energy crops", *i.e.* crops for ethanol or biodeisel, could offset 300-1300 million metric tons of carbon (or 1100-4800 metric tons of CO_2) per year from fossil fuels, by capturing previously emitted carbon and burning it again. Simply turning crop residue (*i.e.* stalks of plants, straw, etc.) into biogas, or direct burning it, could contribute 100-200 million tons of carbon (400-700 Mt CO_2), without disrupting food production .

ENSURING GLOBAL FOOD SECURITY

There are many ways to affect climate change and secure our future food supply. However, the future is very uncertain. The IPCC makes recommendations, sometimes contradictory, but in the end, much of this is new ground. The IPCC recommendation is to use agriculture for producing biogas, but widespread production of ethanol in developed countries, along with incentives by national governments to produce ethanol, has been discovered to be a major contributor to the large rise in food prices in 2007 and 2008 by World Bank Economists and others. For reasons probably unrelated to Climate Change, China is undergoing severe agricultural decline.

They have lost one fifth of their agricultural land since 1949 due to soil erosion and economic development (CIA, 2008). China's imports of grain are growing as their population grows and environmental problems impact agriculture, and they are projected to have a grain production deficit on the order 40 million tons per annum. With China and other industrialized nations with grain deficits hunting for scarce food exports, the rise in prices is hitting poorer nations particularly hard. Countries with grain surpluses have been halting exports to guarantee food to their own populaces, as surplus food becomes scarcer. African nations are among the hardest hit; Mauritania has seen a 67% increase in the cost of wheat, a 117% increase in the cost of cooking oil, and a 25% increase in the cost of rice (the three key ingredients for food in the nation).

Mauritania currently has a roughly 45 day supply of these commodities; and producing nations are being stingy with imports. Wealthy importers such as Malaysia and Singapore are hording food and banking against even higher prices. People are starting to leave their homes in search of work and food. Even the United States is feeling the crunch; the price of wheat flour has doubled in just a few months, probably linked to a 24 million acre drop in wheat planting since 1981; some of which is due to the production of ethanol, and corn and soy for export. With all of these shortages happening now, we can only wonder what climate change is going to do to an already dire situation. It may be true that climate change is already be affecting production in a limited way, but by all accounts, it will get far worse despite our best efforts. What will the world look like then?Food supplies and poverty levels have been significant global issues for centuries. The concept of achieving total food security, defined as "access by all people at all times to enough food for an active, healthy life," is one of the most challenging goals for humanity. However, with the recent onset of global climate change, providing enough food to sustain human populations is becoming increasingly difficult. Most models show that global warming will reduce agricultural output, leading to shortages in many developing countries. With less supply come higher costs, and in turn less availability to already impoverished regions. Studies show that the number of hungry people in developing countries will increase by 1% for every 2-2.5% increase in prices. In 1980, there were an estimated 500 million people at risk of hunger. In 2060 this number is projected to be around 640 million, ignoring the effects of global change. The increase in cost due to lower yields is expected to increase this statistic by anywhere from 60 to 350 million people.

For researchers modeling the impact of climate change on poverty levels, agriculture has been the most significant variable. This is because many of the staple crops used to feed human populations are naturally more vulnerable to climate shifts. Studies have predicted agriculture's role in reducing poverty will decrease, leading to around 600 million additional people with inadequate food supplies. There is some debate as to whether small or large-scale farming would adapt better to global warming. While small farming are generally thought to be good adapters, they would not fare as well long term in a shifting environment, as they cannot afford to take risks.

In creating a viable and enduring agricultural system for the 21st century and beyond, great care must be taken to craft agricultural systems that take into account the dual problems of increasing population and food demand, and climate change.

The problems are multivariable, interrelated and profound; and there is no clear answer facing us at this time. But at the same time, these problems are not going away, and much work and research is still needed to see us through.

Though most research considers a growing population a given of these agricultural problems, we feel that governments singularly and as a whole should begin to consider widespread population control measures.

Population is growing, especially in developing countries, some of which are the hardest hit by the growing food crisis, and curtailing growth would reduce much of the strain to continue expanding agricultural production in the face of climate change and adverse effects such as deforestation. We believe that by implementing moderately strict population control along with these solution and mitigation practices, we stand a good chance of being able to feed our people for the next century. Humans have always faced the problem of feeding themselves, but never has the scope of the situation been so wide, and the solutions so grand and inspiring.

Conservation vs Food Production: a False Dichotomy

It is tempting, given the question's use of the term "conserving ecosystems," to focus this answer on the 14% of the terrestrial system that is protected. Such a response, however, would miss the elephant in the room. To conserve ecosystems in a changing climate, we must instead focus on how they are used and managed in the remaining 86% of land cover. Food production currently dominates that use, with 40 percent of the Earth's land cover managed for crop production and pasture.

To date, the focus on managing ecosystems for food production has resulted in significant benefits for humans. The Millennium Ecosystem Assessment found that globally human well-being has improved over the past 50 years, as measured by the UN human development index. We live longer, are better nourished, literacy rates have risen along with educational attainment, and per capita GDP is rising.

Yet, as human well-being has improved, the health of ecosystems has declined. Humans have degraded the majority of ecosystem services""capture fisheries, freshwater, pollination and many more""while enhancing a few food services. Food production is a major force behind every direct driver of ecosystem degradation identified by the Millennium Ecosystem Assessment in 2005.

Food production as a major driver of ecosystem degradation Habitat conversion: Approximately 43% of tropical and subtropical forests and 45% of temperate forests have been converted to croplands. Over exploitation: 70% of global freshwater use is by agriculture, reducing its availability for other uses.

Invasive species: The introduction of aquatic alien fish species has led to the extinction of native species in many parts of the world.

Pollution: Only a fraction of nitrogen applied as a fertilizer is typically used by plants, the rest ends up in inland waters and coastal systems, creating eutrophication and dead zones.

Climate Change: Agriculture directly contributed to around 14% of global greenhouse gas emissions in 2005 and drives additional emissions through its role in deforestation.

Minqin Oasis in Western China, for example, historically served as a natural barrier against the dryness of the Tengger and Badain Jaran deserts. In the 1950s Chairman Mao implemented a national plan to boost food production, entailing cultivation, deforestation, irrigation, and reclamation. The long-term consequences on other ecosystem services such as erosion control and freshwater were devastating. As a result, the Minqin oasis has been swallowed by deserts, leading to abandonment of once productive land. In the U.S. midwest region, the production of crops, livestock, and biofuel has degraded coastal ecosystems. In the Gulf of Mexico, nutrient runoff from farmland has created a dead zone approximately the size of New Jersey which has degraded fisheries and is expected to worsen with climate change.

Can this paradox of improving human well-being by increasing food production at the expense of other ecosystem services that humans depend on continue? Scientists worry that time lags may exist between the degradation of ecosystems and the resulting effects on human well-being. And that these may lead to tipping points past which damage to ecosystems is irreversible. The Millennium Ecosystem Assessment identified several examples of irreversible ecosystem change, such as the sudden collapse of chronically over-exploited North Atlantic cod stocks in the late 1980s. As climate change impacts exacerbate food production stresses on ecosystems, it is conceivable that such collapses become commonplace, with enormous implications for food security, especially in the developing world where two billion rural poor depend on ecosystems for sustenance and livelihoods. For example, the Amazon provides critical water and climate regulation services that the region's agricultural sector depends upon for its survival. Yet scientists fear that ongoing deforestation combined with climate change could bring the Brazilian Amazon to a tipping point beyond which it experiences widespread dieback and transitions into savanna-like vegetation. The reductions in rainfall would devastate efforts to raise crops and cattle in the region.

Further upping the food production/ecosystems health challenge facing policymakers, population growth and rising per capita incomes are expected to double the world's demand for food in the next 40 years, according to UN food and agriculture chief, Jacques Diouf.

Avoiding Ecosystem Services Tipping Points

To conserve ecosystems in the face of climate change, confronting the elephant-the way we meet our needs for food-must be a prime focus. And our approach to conserving ecosystems must ensure that they are not pushed beyond dangerous tipping points. Future approaches to conserving ecosystems must tackle the three interlinked challenges of climate change, ecosystem services degradation and rising demand for food.

Climate change : Climate change affects the quantity and quality of food and other ecosystem services. Moderate warming is expected to benefit agriculture in mid to high latitudes, but to decrease yield of cereals in low latitudes (medium confidence). The loss of ecosystem services is likely to be a major source of surprises from climate change. While ecosystems have adapted before in response to shifts in climate, their resilience this time is undermined by existing and growing human pressures. Future food production must therefore reduce its pressure on ecosystems and plan for the unexpected, such as changes to freshwater, natural pest regulation and pollination services.

Ecosystem service degradation : The majority of ecosystem services globally have been degraded, including those that underpin food production. Crops, livestock and aquaculture are among the few ecosystem services that have been enhanced. The degradation of ecosystems services is expected to grow significantly worse in the first half of the 21st century, with climate change possibly becoming the dominant driver by the end of the century.

Doubling of Food Demand

The focus on increasing food production needs to continue, if we are to sustain previous gains in human wellbeing, feed a growing population and lift millions out of poverty. But increasing food production at the expense of other ecosystem services and climate change cannot be sustained indefinitely. Approximately 13 million ha of forest, for example, are lost annually, primarily as a result of the expansion of land used for food and fuel. Deforestation is a significant cause of climate change and results in the loss of ecosystems services that are critical inputs to agriculture, including erosion control, climate regulation, pollination and water regulation. Yet, projections estimate conversions of an additional 120-240 million ha by 2030, much of it in Latin America and Sub-Saharan Africa and much of it for food production.

Addressing the intertwined challenge of meeting rising demand for food and conserving ecosystems in a changing climate will not be easy. It will require policy makers at all levels of society adopting different approaches to conserving and managing ecosystems that factor in climate change and other drivers of ecosystem change. Two such approaches are described below: 1. integrate climate change and ecosystem service risks in decisions that affect ecosystems and 2. reconcile food production and conservation goals.

Integrate climate change and ecosystem service risks in decisions. Future ecosystem use decisions, particularly those that involve food production, should assess the risks associated with their dependence and impact on ecosystem services and how these risks will be altered by climate change. A rapid assessment tool for this purpose, based on the Millennium Ecosystem Assessment conceptual framework and the World Resources Institute's subsequent work.

This comprehensive tool identifies the direct drivers of ecosystem change, including climate change, in a particular setting, and assesses both risks and opportunities stemming from a decision's dependence and impact on ecosystem services. It can be applied by policy makers working in different sectors and at multiple scales, from field to forest or watershed from to national to regional and local levels.

Reconcile food production and conservation goals. Future approaches to food production must first and foremost recognize its interdependence on a basket of ecosystem services and develop strategies to sustain them in the face of climate change. To achieve this, food security experts need to work alongside agriculturists and biologists to maximize food production while minimizing further ecosystem and biodiversity loss. Three key strategies can help meet this goal:

Restore Degraded Lands

There is an urgent need to halt the expansion of food production into natural ecosystems in order to conserve ecosystem services and biodiversity. Globally, over one billion hectares of land is believed to have restoration potential for increased human use-a global combined area greater than China. More work is needed to determine which of this is suitable for food production, taking into account water availability, climate change and social factors. Restoring even a small part of this for food production would help reduce pressure on natural ecosystems. In Indonesia, for example, the World Resources Institute is seeking to develop a scalable model for diverting new oil palm plantations that would otherwise replace virgin forests onto degraded land. This ensures that oil palm plantations can keep expanding to meet demand""generating local revenues and jobs while reducing deforestation. Similar opportunities exist to divert the expansion of cattle ranches in the Amazon region to degraded lands.

Increase Productivity on Existing Farmland

While intensification doesn't immediately come to mind when thinking about conservation, it is nevertheless a key strategy to reduce stress on natural ecosystems. The challenge is to find ways to produce more food without unwanted ecosystem trade-offs. As noted by Shivaji Pandey, Director of FAO's Plant Production and Protection Division at a 2009 Congress on Conservation Agriculture in New Delhi. "In the name of intensification in many places around the world, farmers over-ploughed, over-fertilized, over-irrigated, over-applied pesticides. In doing so we also affected all aspects of the soil, water, land, biodiversity and the services provided by an intact ecosystem. That began to bring yield growth rates down." Food production needs to be redesigned to be both more productive and synergistic with other ecosystem services. Public subsidies and tax incentives, for example, should be used to compensate farmers for managing their lands in ways that sustain beneficial ecosystem services to society. To contend with climate change, farmers need location-specific information on ecosystem services and changing climate conditions. This should be combined with research on how improved practices can address climate constraints on crops. For example, rainfall may remain plentiful in parts of Africa but be concentrated in shorter time periods. Ways to minimize run-off and soil erosion and store water to extend growing seasons will thus be important.

2

Financing of Agricultural and Food Supply Chain

FINANCING-FOOD PROCESSING SECTOR

The food processing sector comprises a large number of small and medium sized companies, a significant proportion of which have stand alone operations, with no control over the raw material base and reliant on other organizations to undertake marketing/further processing of their products. Banks and financial Institutions adopt the same risk models relevant to the manufacturing sector, for assessing food processing companies. Interest charges for food processing companies are high, on account of the high risk perception associated with the nature of their operations.

Food Processors' Perspective

With raw material availability often being seasonal, and/or concentration in demand, inventory holdings are high and thus the working capital finance issued through normal Maximum Permissible Bank Finance method does not address the funding requirements adequately. The rate of interest for working capital assistance is high. Most food processing companies are not able to access adequate working capital at reasonable rates. This affects the raw material procurement and capacity utilization through out the year.

Unavailability of large working capital facilities, or facilities without a high seasonal drawal, restricts players in purchasing agri produce in large volumes, when prices are favourable.

Banker's Perspective

- Lack of reliable information on demand-supply, price trends, raw material supplies etc is a significant constraint for banks, impacting credit assessment and monitoring

- High operational/transaction cost in servicing small companies
- Lack of backward linkages for assured access to raw material
- Lack of forward linkages of stand-alone food processing units for marketing/distribution

Owing to high cost of market development, the profitability of food processing companies is under pressure in the initial years of their operation. This impacts the risk rating and leads to high cost of borrowing.

Even temporary adversities in market conditions can lead to defaults and eventually the loan can become a non performing asset for banks.

FINANCING TO AGRICULTURE AND FOOD PROCESSING

In order to improve the funding from the institutional set-up to agriculture and food processing, the following policy changes are required:

Restructuring of Government Schemes

- In capital intensive projects, the interest can be capitalized and funded, so that the burden of high cost of interest is not faced by the company in the formative years
- The agencies could structure schemes in the form of capital subsidies, which in turn could be leveraged by the companies to access bank finance. Release of funds should be proportionate to the equity brought in by the promoters.
- In sectors, which are working capital intensive, the funding institutions should consider offering soft loans instead of grants. The soft loan would ensure that the companies undertake operations to service the interest payable to the institution.
- Market analysis prior to providing funding-Detailed market analysis of demand-supply trends, costs, competitive position should be undertaken by the funding institution/by a neutral third party which has the required industry expertise, to ascertain the financial viability of the project
- In order to address the issues related to food parks, the recommendations are as follows:
- It is recommended that food parks be located in proximity to raw material sources to maximize the potential available locally. This will ensure better utilization of common facilities.
- The quantum of financial assistance to a food park needs to be decided on a case-specific basis, depending on the infrastructure requirements

across locations. The financial appraisal of any food park scheme should be evaluated on the basis of potential for tie-ups with off takers and sufficient availability of raw materials.

- The definition of common facilities could be expanded to include roads, drainage and solid waste management facilities etc.
- It is critical to have a monitoring mechanism to facilitate and review progress of implementation. The State Departments for food processing can play an active role in this process.

Redefine role of Regional Rural Banks (RRBs)

- The RRB Act needs to be amended to permit RRBs to close branches and allow merger/amalgamation of RRBs among themselves or with the sponsor banks to take advantage of operational synergies. The capital structure of the RRBs also needs to be revised since state governments have been unwilling to inject their share of capital in past recapitalization exercises. The government may permit sponsor banks to increase their stake in the RRBs or bring in new investors. With this reorientation, RRBs can be effectively utilized by banks and Financial Institutions as channels for rural lending. B. Amendments in Priority sector Lending
- Definition of priority sector: In a recent circular, Reserve Bank of India has imposed a cap on investment in plant and machinery upto INR 50 million for food processing industries for classification under the priority sector. As mentioned in earlier sections of the report, as also detailed in subsequent sectoral sections, for Indian food companies to compete effectively, the policy environment needs to address all bottlenecks hindering scale of operations, and availability of financing features high on this list.
- All food processing units should be classified as priority sector irrespective of size. Relaxation of the cap on investment for classification as priority sector would provide the required fillip for flow of credit to the agribusiness sector and would also have a beneficial impact on farmers.
- Further, the scope of priority sector lending should be increased to include all primary processors of agriculture produce under 'priority sector' category, including sugar manufacturing units, rice mills, dal mills etc.
- Farmer financing against warehousing of agricultural produce: Priority sector norms stipulate a cap on advances against warehouse receipts of produce upto INR 0.5 million. The norm also stipulates the advance to be exclusive to farmers availing a crop loan with the same Bank This

cap on the quantum of borrowing should be removed. Further, the advance should be delinked from the crop loan. This will encourage farmers to store their produce during the peak season and sell it during the lean period, to obtain better prices for their produce. The current system forces the farmer to sell all his produce immediately after harvest to traders/commission agents, who store the produce and profit due to price fluctuations.

- Financing for high value crops: Cultivation of crops such as grapes requires high upfront investments, estimated at INR 0.8 1.0 million per acre for establishing of vineyards. The cap on farmer loans for a maximum amount of INR 0.5 million per farmer restricts bank finance. This cap on farmer loans should be made category/crop specific, taking into account the respective cultivation and investment expenses.
- At present, banks typically provide financial assistance through loans for a period of 3-4 years. However, infrastructure such as temperature controlled storage have a longer gestation period. There should be a specific focus on encouraging banks to provide long term loans for a period of 5-7 years, which should be classified as direct agriculture lending.
- Lending to agriculture is usually dictated by the scale of finance method of assessment of the credit requirement. This system is outdated as the norms are not based on current costs of cultivation. Banks need to assess credit needs of farmers based on current cultivation costs.
- As mentioned earlier, priority sector lending records the outstanding credit in the books of the banks on the last working Friday of March every year. RBI and NABARD should monitor banks' outstanding portfolio at least twice in a year, so that banks undertake mandatory lending to the agriculture sector during the kharif season as well.

Promote Warehouse Receipt-based Financing

Warehouse receipt finance is a form of secured lending to owners of commodities, (farmers, traders, processors) which are stored in a warehouse, with the commodities having been assigned to a bank through warehouse receipts. On the back of the warehouse receipts being housed with the bank, the latter has the security of goods until they have been sold and the proceeds collected.

Warehouse receipts are a proven tool for financing, utilizing the stored goods as collateral for loans.

The system has been working successfully from the beginning of the twentieth century in several countries such as Brazil, Indonesia, Singapore and

Argentina. In addition to addressing the needs of food processors, as mentioned earlier, farmers can also benefit because it enables them to store produce and sell when prices are favourable.

Some of the advantages of a warehouse receipt system are:

- It provides a choice to primary producers who can decide to sell immediately after harvest or to store in a licensed warehouse and to apply for short-term credit. The farmer can decide to sell his crop later in the year, when prices are typically higher than at harvest time.
- It enables primary processors to purchase raw materials at favourable prices and store until required for processing
- The system leads to a reduction of post-harvest losses as the produce is stored under appropriate conditions in licensed warehouses.
- It translates into lower risks for the financier, as the collateral for the loan is a liquid asset.
- This can lend greater depth to operations of Commodity Exchanges, by increasing the number of transactions without physical movement of goods, through endorsement of the warehouse receipts

Infrastructure & Project finance

Infrastructure development can play an important role in improving efficiency across the supply chain and hence any financial assistance for rural infrastructure development including rural roads, markets, silos etc. should be classified as priority sector. It is recommended that the Government undertake the following measures to boost infrastructure creation for the agriculture and food sector-

- Creating an enabling legal and regulatory frame work for development of infrastructure
- Providing seed capital and leveraging the same for development of agri-infrastructure
- Monitoring and facilitating private sector participation
- Risk sharing with banks in financing agricultural infrastructure projects.

Public Private partnerships are key to development of agricultural infrastructure projects. The linkages between various players, which includes the users, the supply chain participants and operators are crucial. The projects can be structured based on the BOT (and its variant) model which ensures appropriate risk allocation between various players, including the Government.

Some of the project structures which can be adopted are as follows:

- User Charge-based project structure: A Special Purpose Vehicle (SPV), funded by the sponsors and the financier is constituted. The Users, under a waterfall agreement pay on usage of the facility. These user charges service the debt.

- Rated lease-based project structures: The Sponsors assign the assets to a Special Purpose Vehicle (SPV), financed through a mix of debt and equity (from financial investors).The SPV executes lease agreements with the lessee, who in turn pays lease rentals, which are used for regular debt servicing.
- Receivables-based structure: This type of structure can be used for construction of rural roads or market yards, wherein the project is financed based on future receivables from a determined and realizable set of cash flows.

TRANSPLANTING THE SEEDLINGS

This job should be done as soon as the plants raised in the nursery have been hardened off. This involves acclimatizing the plants to field temperatures and usually starts when the young plants are about 3 inches (7-8 cm) high. If the seedlings have not been produced in modules but in boxes it is helpful for the hardening off and field establishment processes if blocks are cut with a knife so that each plant will come out with a root ball when planting later. The field, or plot, should meanwhile have received its final preparation and be ready to receive the plants. A sticky soil surface should be avoided if the plot has to be irrigated before planting otherwise soil structure will be damaged.

A cool time of day should be used for planting. In The Bahamas the afternoon or evening is often considered to be the best time for planting, ideally following rain or irrigation (provided that the soil is not too sticky). The seedling boxes should be watered before planting out, and watered again if any boxes start to dry out before planting is finished. The advantages of this watering are that the plants will come out of their modules with the minimum of damage to the root balls and any immediate wilting after planting as a result of disturbance will be reduced.

The spacing between the rows and within the rows will depend on the vigour and growth habit of individual varieties being grown. The wider spacing is also more commonly used where leaf, stem and fruit diseases are likely to occur. Unless planting on ridges or mounds the young pepper seedlings are planted in rows five feet (approximately 150 cm) apart and three feet (approximately 90 cm) within the rows. This spacing will provide a final plant stand of almost three thousand plants per acre; a spacing of four feet (120 cm) between the rows and three feet (approximately 91 cm) between the rows will give 3 630 plants to the acre; (2.47 acres equal one hectare).

Every care must be taken to keep the root ball of the module unbroken when removing the seedlings from the modules or trays and also when planting them, otherwise undue root damage and wilting will occur which will give the plant a setback or check before it recovers.

The holes to take the young plants should be sufficiently deep and wide to take the root ball or plug without damage. The seedlings should be planted at the same finished depth as they were in their modules, cells or trays; that is to say the finished soil level after planting should be the same as when the plant was growing in its nursery container.

Any plants that show symptoms of virus, stunting or any other undesirable character must be discarded. Virus infected and/or any other diseased plants should not be handled during the planting operation but be disposed of as soon as possible by burning. Some viruses are passed from one plant to another by sap transfer during handling and attention should be given to this point. The newly planted seedlings should either be watered in or irrigated immediately after planting. When planting large areas the newly planted blocks should be irrigated before they start to dry out, especially under drying conditions.

NUTRITION OF THE HOT PEPPER CROP

The basis of pepper plant nutrition is to maintain plants with a balanced growth between those which do not flower or form fruit before the plant has reached sufficient size and plants which remain in a vegetative state by being slower to come into flower than the variety's character.

The main fertilizers applied to the growing crop are nitrogen, phosphorus and potassium. There may also be occasions when another nutrient is required, due to a deficiency of an essential nutrient.

However, advice from an extension officer should be sought before attempting to remedy a suspected nutrient deficiency. In fact advice should sought from extension or advisory officers whenever in doubt about any aspect of crop nutrition; in this way a better understanding will be gained by the farmer.

Fertilizing at planting time: When the base fertilizer has been added during the final stages of soil preparation there is normally no need to apply a nutrient feed immediately after planting. However in areas or seasons of very high rainfall when serious soil nutrient losses from leaching can occur it may be preferable to apply the feed soon after planting. If applying fertilizer after planting it is usual to apply 1 ounce (28 grams) of a general bag fertilizer with an N:P:K ratio of 11:22:22. This should be sprinkled around each plant but taking care not to get any on the plant's leaves otherwise they may become scorched and then become prone to disease. If conditions are dry this fertilizer application should be watered in, but care must be take not to let it be washed off into the gullies.

Fertilizing at the start of flowering: Under normal conditions the plants start flowering at about forty days from planting out. The plants should be given another fertilizer application at about this stage. The aim now is to assist the plant to produce flowers and fruit. Therefore the fertilizer applied has a higher proportion of potassium to nitrogen. A suitable fertilizer has an N:P:K ratio of 8:4:32. This is applied at the rate of half an ounce (14 grams) per plant.

Fertilizing when the first fruit have developed: The pepper plants will start to set fruit within four weeks after the first flowers have been seen. The fruit development stages from flowering. When the plants start to bear fruit another fertilizer application is made. As at the start of flowering stage another application of N:P:K 8:4:32 is applied at the rate of half an ounce (14 grams) to each plant.

Fertilizing after first fruiting: Following the application given when the first fruit were developed further applications are made approximately every four to six weeks. These feeds are a little higher in nitrogen and the N:P:K ratio is 11:22:22 (the same as would have been used for the first feed after planting if required). This provides an increase in both nitrogen and phosphorus so as to maintain satisfactory plant growth and continued development of both roots and leaves. The shorter period (of four weeks) is usually suitable on sandy soils which drain quickly, while the six week period is more suitable for plants on the clay soils.

Methods of fertilizer application: When the fertilizers are applied by hand to individual plants the materials are usually applied in a circle around each plant. An alternative is to apply the fertilizer in a band along the rows of plants. If a serious nutrient shortage has been detected then foliar feeds can be used, these will give a quicker response from the crop plants. But if foliar applications are to be given ensure that the material you plan to use is suitable and be very careful to follow the manufacturer's instructions for dilution, application rate and conditions. Growers who have drip or trickle irrigation systems have the opportunity to apply liquid fertilizers through the irrigation system.

CROP RESIDUE AND SOIL MANAGEMENT

Livestock can enhance farm output by intensifying nutrient and energy cycles, particularly in LEIA and NCA. In LEIA this mainly implies a closer recycling of nutrients in straws (crop residues and fodder), dung and urine. In NCA there is scope to include leys, anti-erosion bunds and, for example, adjusted cropping patterns. NCA also has the added advantage that it can both import some nutrients in fertilizer and feed while it aims to keep the better-quality crop residues such as brans and cakes on the farm. Crop residues can be used for animal feed as well as for soil fertility and can also be sold. The farmers have to choose how to use these materials and, whatever they are used for, the choice needs to be accompanied with keen management to make the best of these resources.

The fibrous crop residues have a potential in soil conservation through mulching but even the oilseed cakes can be used to improve soils. Some terms related to animal-plant-soil interaction.

ELIMINATING THE EFFECT OF FOOD PRODUCTION ON THE CLIMATE

Agriculture accounts for between 10-12% of total human greenhouse gas emissions, with 60% of methane (CH_4) emissions, and 50% of all human nitrous oxide (N_2O) emissions. While food production's share of greenhouse gasses is small overall, its share of both nitrous oxide and methane is large. These gasses have a far higher effect than carbon dioxide which is by far the most prevalent gas.

Methane is about 21 times more effective at trapping heat than CO_2 (EPA, 2007), and nitrous oxide is far more than that, being 310 times more effective than CO_2 (EPA, 2006). From this, we can see that the effect of food production, while by no means the single source of Global Warming Potential (GWP), has a major effect. In addition to adapting our agricultural system to continue to provide us with secure sources of food in a warmer world, we must also adjust our agricultural practices and implement technology to reduce the affect of food production on the problem of Climate Change itself. Mitigation practices fall under several categories; reducing emissions, enhancing removals, and avoiding (or displacing) emissions. The first category involves changing practices to put less GHG's into the atmosphere. Changing feed practices for livestock and altering the method of nitrogen application to crops are two examples of practices that would reduce emissions of methane and nitrous oxide respectively.

Enhancing removals deals with improving the carbon sequestering potential of food production. According to the IPCC's Third Assesment report (2001), agricultural ecosystems have a high carbon storage potential, mostly in organic matter within the soil.

Processes that increase photosynthesis and slow re-release of carbon via plant respiration or erosion will affect this result. Such stores of carbon are referred to as "carbon sinks". While sequestering other gasses is possible on a more limited level, the primary sequestering potential for CO_2- is sometimes at a cost of more emissions of other GHG's; such as methane or nitrous oxide. Once again, the methods and technologies used are expected to vary widely by local climate and geographical conditions.

The final method, avoiding or displacing emissions, involves using agricultural bi-products and crops as a source of energy and replacing fossil from fossil fuels or other sources.

These "biofeedback" energy sources still release GHG's, however they gain benefit from the fact that the carbon used was recently removed from that atmosphere (an active part of the carbon cycle), and not from fossil fuels, where it was stored underground for centuries.

With this sort of energy source, the only 'new' carbon introduced to the cycle is that used to produce and process the energy source. Other options, such as using methane from animal waste products to power generators, takes methane that would otherwise be emitted directly to the atmosphere and uses it to generate electricity.

Reducing emissions from croplands, especially of methane and nitrous oxide, can often be effected by altering the way the crops are managed. Crop management can also effect how much carbon is taken in by the plants. Adding nutrients to the soil, for example, can improve the carbon gaining potential of the plants. But too much fertilizer can be a bad thing, increasing the emissions of N_2O. Precise regulation of fertilizer application can greatly reduce excess N_2O emissions; as can strict control of leaching. Periodically draining rice paddies could help control high methane emissions.

Unfortunately, this could cause a spike in N_2O production from those same fields. Many of these practices have tradeoffs; either in terms of increased emission of another greenhouse gas, or damage to ecosystems in other ways.

Loss of forests (a better carbon sink than croplands), due to increased agricultural production would clearly be a bad thing; while increased efficiency of production on current cropland (with possible spike in emissions) could slow the inevitable growth of agricultural lands, and prevent emission-high deforestation. Implementation of all of these practices must consider these factors and many more. A long view is needed; many of these practices change over time.

FACTORS AFFECTING MICROBIAL BEHAVIOR IN FOODS

Most studies in food microbiology are concerned with the rapid growth of populations, but in many ecosystems, the survival characteristics of the population also need to be considered. The longevity of bacterial spores and their resistance to harsh conditions are well documented. However, the ability of vegetative cells to resist stressful conditions is increasingly recognized as an important ecologic trait. Attention also needs to be given to relatively slow-growing populations in various situations, *e.g.*, when the shelf life of a product is extended by control of rapidly growing spoilage organisms. The behavior of foodborne microorganisms, be it the growth or death of microbial populations, is based on the time of exposure to environmental factors affecting population development; for example, equivalent kills of bacteria in milk are achieved by low temperature-long time pasteurization (60°C/30 min) and high temperature-short time pasteurization (72°C/15 sec). When populations are in the biokinetic range, the rate at which they develop is determined by factors such as temperature, water availability, and pH applied in food preservation procedures. The extent of microbial growth is a function of the time the population is exposed to combinations of intrinsic food properties (*e.g.*, salt concentration and acidity) and extrinsic storage conditions (*e.g.*, temperature, relative humidity, and gaseous atmosphere).

Different factors assume dominance in different foods and preservation strategies. In many foods, the full preservation potential of a single property is restricted because of considerations related to the esthetic, organoleptic, and nutritional properties of the product. However, several properties or conditions may be combined to provide a desired level of stability. In situations where the preservation strategy is designed to slow the rate of population growth, the effect will always be increased by storage temperature.

Temperature control in processing, distribution, and storage (the cold chain) is crucial to ensure the adequate shelf life and safety of many common foods, including meat, fish, poultry, and milk. Newer technologies, including modified atmosphere packaging and sophisticated products such as sous-vide meals, do not obviate the need for strict temperature control. Indeed, the requirement for vigilance increases with increased shelf life and the possibility of growth of psychrotrophic pathogens over an extended period.

Predictive Microbiology

Predictive microbiology involves knowledge of microbial growth responses to environmental factors summarized as equations or mathematical models. The raw data and models may be stored in a database from which the information can be retrieved and used to interpret the effect of processing and distribution practices on microbial proliferation. Coupled with information on environmental history during processing and storage, predictive microbiology provides precision in making decisions on the microbiologic safety and quality of foods. The term "quantitative microbial ecology" has been suggested as an alternative to "predictive microbiology."

The development, validation, and application of predictive microbiology has been extensively reviewed in the last decade. Modeling studies have concentrated on descriptions of the effect of constraints on microbial growth (rather than survival or death), often using a kinetic model approach (rather than probability modeling) and most often describing the effect of temperature as the sole or one of a number of controlling factors. For example, the temperature dependence model for growth of Clostridium botulinum demonstrated a good fit to data, but the authors noted that "care must be taken at extremes of growth, as no growth may be registered in a situation where growth is indeed possible but has a low probability."

The emphasis in modeling efforts on temperature (often in combination with other factors) may be justified, given its crucial role in the safe distribution and storage of foods. Surveys carried out over several decades in the United Kingdom, United States, Canada, and Australia point to the predominant role of temperature abuse in outbreaks of foodborne disease.

Problems with Predictive Microbiology and Research Needs

Several commonly perceived problems with predictive modeling are reviewed below. While considerable progress has been made in defining philosophic approaches and experimental protocols for growth model development and many models have been developed and published, more validation studies are required, particularly involving independent and industry-based trials. More emphasis should be placed on modeling the death kinetics of foodborne pathogens with low infective doses.

Measurement of environmental factors (*e.g.*, temperature) can be achieved with precision, but in some situations, (*e.g.*, in chilling of meat carcasses), it is more difficult. Location of the sensor can be an important consideration.

In meat chilling, where control of microbial development is a function of the combined effects of falling temperature and water activity, development of a technique to measure water activity in situ at the carcass surface would provide valuable information. Furthermore, development of techniques to measure constraints such as water activity, pH, or redox potential on a microscale might provide useful information for a complex food such as salami. This would allow definition of the role of the microenvironment in determining microbial behavior. The concept of a microenvironment is well developed in soil microbiology but has been neglected in food microbiology.

The inherent variability of response times (generation time and lag phase duration) as an issue in predictive microbiology was first raised by Ratkowsky et al., who related the variance of responses on a transformed rate scale such as V(Ök) or V(lnk) to the variances of responses on a time (q) scale. The variance was shown to be proportional to the square or cube of the response time. It was later confirmed that nonnormal gamma and inverse Gaussian distributions described the distribution of response times in predictive microbiology, which indicate that variance is proportional to the square or cube of the response time, respectively.

The practical implication of these findings for the application of kinetic models is that inherent biologic variability increases markedly with increasing response times, and thus the confidence limits associated with predictions also increase markedly. However, if the probability distribution of the response time is known, one can determine the probability that an organism will grow more quickly than a predicted response time. Thus, kinetic models are appropriate to describe consistent microbial growth responses, but under extreme conditions a probability approach may be required. Models are normally developed under static conditions (growth rates and lag times are measured at a series of set temperatures, water activity values, and pH levels), and the results are combined to describe the effects of each factor or a combination of factors on population development. Subsequently, models must be validated in foods under conditions that mimic situations encountered in normal practice, *e.g.*, decreasing temperature and water activity during active chilling of meat carcasses or fluctuating temperatures during the distribution and storage of many food commodities.

Shaw and later several other authors reported on the effect of fluctuating temperatures. Depending on the magnitude of the temperature deviation, the organism may change its rate of growth to a rate characteristic of the new temperature, or it may stop growing if a lag phase is introduced. In both situations, Salmonella Typhimurium has responded nearly as predicted by the model. Baranyi et al. presented similar results for the spoilage bacterium Brocothrix thermosphacta. When cycled between 25°C and 5°C, the model predicted behavior well in both the growth rate and lag phase duration. However, a temperature shift from 25°C to 3°C caused deviations from model predictions due to decline in viable cell numbers or extended lag phases.

During the final extended phase of growth at 2.8°C, the rate resumed at that predicted. Baranyi et al. also examined the perceived problem of modeling lag phase duration. The difficulty in predicting lag phase duration in foods is not due to the lack of a suitable model: the difficulty comes from the lack of knowledge of the physiologic status of the organisms contaminating the food. The organisms may include cells that are actively growing, exhibiting a physiologic lag phase, damaged and under repair, exhibiting physiologic (endospores) or exogenous dormancy (VNC cells), damaged but unable to reproduce because of ineffective repair mechanisms, and dead. At least part of the confusion surrounding the measurement of lag phase duration arises from experiments in which inocula of different physiologic statuses were used.

Methods to define the physiologic status of foodborne contaminants under various conditions need to be developed. This will require observations on individual cells or small populations of cells either directly by microscopy or an indicator of single-cell metabolic activity. Luminescent Salmonella strains have been used as real-time reporters of growth and recovery from sublethal injury. Alternatively, a parameter to describe the suitability of cells to grow in a new environment may be incorporated in the model.

FOOD RETAILING ON THE SUPPLY CHAIN

Organised food retailing in China has increased from less than a billion dollars in 1994 to $55 billion in 2002.ln the corresponding period, the number of organised retail outlets has increased from 2500 to about 53,100. Organised retailing has over a 33% share of the urban food market and 11% share of the total food retail market in China.

This development has been driven by factors which have also been observed in other developing countries (urbanization, income growth, and liberalization of foreign direct investment in retailing) as well as specific policies of the Government (government investment in the sector, and policies promoting conversion of wet markets to supermarkets).

China has followed a policy of opening up of the retail sector to foreign players in a phased manner. In China, modern retail grew at 40% after FDI was allowed in 1992.Initially, FDI was allowed only in 11 cities and the number of foreign retailers in each city was restricted. However, these restrictions have been relaxed in the last decade. China has lifted restrictions on foreign investment in retailing, including those on foreign ownership and the number of outlets.

STATUS OF FOOD DISTRIBUTION IN INDIA

Food accounts for the largest share of consumer spending. Food and food products account for about 53% of the value of final private consumption estimated at INR 8600 bn (2003-04 at current prices). Food products are sold through over 5 million food and grocery stores in India.

The size of the organised food retail segment in India is estimated at INR 20 bn (approximately 0.2 % of food expenditure).This is in contrast to most of the developed countries, where food distribution is highly consolidated. In countries such as Australia, France, Germany, Denmark and The Netherlands, the share of top the 10 retailers is more than 80% (USA ~ 35%, China ~ 15%).

Most organised retailers in India are regional and use single formats such as convenience stores, supermarkets, hypermarkets or cash and carry. In contrast, most international retailers have multiple format models. Besides reflecting the stage of evolution of retailing in India, this also highlights the capital constraints of existing players in India, restricting their ability to make large investments.

In India, food companies are much larger in size than the organised food retailers in India. In contrast, the retailers are larger in scale, than food companies, in developed countries. Given the channel power of retailers in these markets, food companies have started establishing partnerships with them to develop and test new products, share consumer information and undertake consumer promotions.

Key Drivers and Impediments for Growth of Organised Food Retailing

Increasing need for convenience: The Indian consumer visits about eight to ten outlets to purchase various food products which constitute the daily consumption basket. These outlets include neighbourhood kirana stores, bakeries, fruit and vegetable outlets, dairy booths and chakkies (small flour mills). With changing lifestyles, there is growing paucity of time, and convenience in food shopping is emerging as an important driver of growth of one-stop retail formats.

Availability of quality retail space: Until the late 1990s, the high cost of real estate meant that organised food retail business models were not financially viable in metropolitan areas.

In the last few years, various factors have led to increased availability of real estate for organised retail formats. About 300 malls are at various stages of construction, across metros and mini-metros in the country. The average size of a mall is about 100,000 sq.ft. About 25 malls each are planned in Delhi and Mumbai, primarily in suburban and satellite areas. This will translate into additional retail space of 30 to 40 million sq.ft over the next three to five years.

Key Impediments

The key impediments to organised food retailing include lack of infrastructure, technology and capital. Lack of infrastructure and technology: Food distribution in India is characterised by a high degree of complexity given dispersed production on the one hand, and variance in demand across locations on the other. There is a compelling requirement for appropriate infrastructure for storage and transportation such as temperature controlled warehouses and vans. However, the limited scale of operations of retailers has restricted their investment capacity in these areas. The recent entrants who have fuelled growth of organised formats are attempting to address this issue, but given the large requirements of capital, the process of transformation is gradual. Further, entry of organised retailers, who have larger investment capacity, will also address the issue of creation of infrastructure for storage and transportation.

Lack of capital: Organised food retailing is a capital intensive business with a long gestation period (typically 5-7 years). Investment is required in buying/leasing land, furniture and fixtures, IT systems, quality control systems, vendor development and infrastructure for warehousing, storage and transportation. The development of organised food retailing in India has been constrained due to lack of capital. Foreign Direct Investment (FDI) is not allowed in retailing with the exception of cash & carry formats.

SUPPLY CHAIN ALIGNMENT WITH INTERNATIONAL REQUIREMENTS

The supply chain needs to be aligned with the requirements of importing countries which require control and monitoring of quality standards of the raw material and processed products. The specific action steps to facilitate this are:

- Enable direct farmer-processor linkages by amendment of the APMC Act
- Set-up independent world-class food testing and inspection infrastructure, particularly in clusters with significant presence of exporters

- Devise an alternate system of processing-grade product specifications based on internationally accepted norms, delinked from fair average quality of table grade products.
- Encourage investment in infrastructure to improve product quality, through part-funding these investments, such as financing of bulk coolers
- Support private sector initiatives for investing in specialized transport infrastructure such as reefer vans through specific financing schemes for this purpose

Integration of Government Schemes

The Government, through various Ministries and allied agencies, offers support to exporters through various schemes to part-finance specific investment requirements.The Ministries/allied agencies include Ministry of Agriculture, Ministry of Food Processing, APEDA, MPEDA, Coffee Board, Tea Board, Export Inspection Council etc. It is essential to align the various offerings of the Government, to address various requirements of exporters and avoid duplication of efforts.

Supply chain issues have been identified as a key constraining factor for exports. The current approach to supporting food/agri exports, through isolated schemes operated by various Ministries/Departments of the Government has not been able to address bottlenecks to exports in an effective manner. There are significant overlaps, while at the same time several need gaps which have not been addressed. Further, the quantum of assistance has a low ceiling, which is one of the factors which contributes to small scale of operations of exporters.

The action plan is stated below:

- Integrate all schemes offered for export promotion through various Ministries and allied agencies such as APEDA, MPEDA, Coffee Board, Tea Board, Export Inspection Council, Ministry of Agriculture, Ministry of Food Processing etc. under one body.
- Strengthen food processing infrastructure in AEZs
- Encourage food testing laboratories in India to get accreditation from international agencies.
- Set-up independent world-class food testing and inspection infrastructure, particularly in clusters, with significant presence of exporters
- Devise an alternate system of processing grade product specifications based on internationally accepted norms.

- Promote aggregation of exports to meet the minimum order requirement of importers
- Expand the list of export products for certification by EIC
- Develop a strong market intelligence system to aid exporters to take rational decisions.
- Introduce certification zoning systems pesticide-free zones, organic production zones, disease free zones to facilitate high value exports from India
- Promote certification for organic farming for different crops
- Build global brands on the back of India's strengths (Darjeeling tea, Basmati rice, Durum wheat, Alphonso mango)

Implementation of Warehouse Receipt Financing

The legal environment must ensure ease in enforceability of the security and the banks must be comfortable to lend against warehouse receipts.

The law should define the following aspects in detail:

- The quality and quantity of the goods stored (credibility of the receipt)
- The rights, liabilities and duties of each party to the warehouse receipt must be clearly defined
- Receipts must be clearly transferable by delivery or endorsement
- If the warehouse owner defaults or his business is liquidated, the holder of the receipts must have the right to receive the stored goods or the financial equivalent.
- Negotiability of warehouse receipts-the status of a pledge is unclear under Indian law, and in the event of a borrower's insolvency, the other creditors will enjoy a prior claim over the receipt holder.
- Uniformity of warehouse receipts across States
- Netting of fungible goods across locations.

Reliable Warehouses at Appropriate Locations

The warehouses should have appropriate facilities for storage such that the grain does not deteriorate. Further, if there are multiple users of the warehouse, the facilities should be such that there is no mixing of the produce. They should be professionally managed and be upgraded periodically. There is a need for development of adequate and quality warehousing systems in India. A good inspection system for the participating warehouses is an important precondition. Due to the fact that the issued warehouse receipts will be treated by all parties as an asset of high liquidity and value, the inspections have to be rigorous to ensure that the predetermined minimum acceptable standards for the warehouse are met. The inspection shall not only be focused upon the technical standards, but also on the financial strength of the operator, the ability to store according to quality standards, and administrative capabilities to ensure that management of the warehouse fulfills the required administrative procedures. A monitoring system with frequent site visits is required. In some countries, State Licensing and Inspection Agencies execute this task while in others, private, independent national or international companies provide these services.

Financiers Involvement

International experiences with warehouse receipt based financing highlight the importance of involvement of strong local banks, as early as possible in the implementation of the program. Their staff has to be trained and clear procedures for evaluation of the commodity must be laid out. In order to reduce the risks associated with commodities, most banks will lend a certain percentage of the market value of the commodity stored, which in the initial years can be 50-60% of market value.

Strong Support of Public Authorities

Strong support of the public authorities is required to set up an efficient warehouse receipt financing system. Their assistance is needed to establish the institutional framework, which would include the following aspects:

- a Assist in setting up of the licensing and monitoring system.
- b Facilitate direct processor farmer linkages (rather than via the mandi) to enable warehouse receipt based financing
- c Dematerialisation of the warehouse receipt to enable electronic trading
- d Development of secondary market for warehouse receipt trade

Well Trained Participants

Participants need to have in-depth understanding of the operations of warehouse receipt systems, including procedures and risk factors. The scale of operations needs to be large, or small farmers need to form associations/groups (which should be a legal entity) to pool the produce, thus lowering transaction costs on the one hand, as also allowing processors access to larger lots of produce.

Certification of Commodities

Grading and standardisation of commodities is essential due to the existence of a large number of varieties which renders price discovery difficult.

Credible certifying organizations who can undertake the task of verification of the commodities stored are required.

Sales Tax: Imposition of sales taxes is an impediment to warehouse receipts. Such taxes are chargeable on goods changing hands in warehouses, with taxes payable on each transaction. These taxes are often higher than the total gross margin earned by market intermediaries.

Commodity Selection

The commodity to be financed under this kind of structure needs to meet the following criteria:

The commodity should be amenable to storage without any significant deterioration in quality or quantity over the tenor of financing. Also there should be a developed and reliable warehousing set-up for the commodity.

This is essential from the lender's perspective because improper storage or any inaccuracies in the warehoused quantity/quality may lead to a fall in security cover.

It may be advantageous if the warehouse providers would be agreeable to providing value added services as may be required by the financing structure.

There should be a developed market for the commodity, providing liquidity without steep transaction costs.

This is essential to provide an avenue for immediate and efficient liquidation of the security if required. In addition, availability of price trends and other trade information about the commodity is essential from the perspective of developing accurate margin norms.

Presence of Derivatives Market

The presence of a derivatives market would be beneficial from a lender's perspectives this will allow hedging of price risk.

Neutral intermediaries are required to be an interface for the lender with the traders. Services that can be provided by the intermediaries include:

- Due diligence of borrowers
- Quality assessment of the commodity
- Historical price and other market information for the commodity
- Faster arbitration and settlement mechanisms

These intermediaries can be service providers, Collection and Management Agencies (CMAs) etc.

FOOD SAFETY

Food safety is a major focus of food microbiology. Pathogenic bacteria, viruses and toxins produced by microorganisms are all possible contaminants of food. However, microorganisms and their products can also be used to combat these pathogenic microbes. Probiotic bacteria, including those that produce bacteriocins, can kill and inhibit pathogens. Alternatively, purified bacteriocins such as nisin can be added directly to food products. Finally, bacteriophages, viruses that only infect bacteria, can be used to kill bacterial pathogens.

Thorough preparation of food, including proper cooking, eliminates most bacteria and viruses. However, toxins produced by contaminants may not be heat-labile, and some are not eliminated by cooking. Source of Contamination should be identified and have eradicate the problem and much more precautions should be taken to eliminate the contamination in foods.

Fermentation

Fermentation is one way microorganisms can change a food. Yeast, especially Saccharomyces cerevisiae, is used to leaven bread, brew beer and make wine. Certain bacteria, including lactic acid bacteria, are used to make yogurt, cheese, hot sauce, pickles, fermented sausages and dishes such as kimchi. A common effect of these fermentations is that the food product is less hospitable to other microorganisms, including pathogens and spoilage-causing microorganisms, thus extending the food's shelf-life.

Food fermentations are ancient technologies that harness microorganisms and their enzymes to improve the human diet. Fermented foods keep better, have enhanced flavours, textures and aromas, and may also possess certain health benefits, including superior digestibility. For vegetarians, fermented foods serve as palatable, protein-rich meat substitutes. Some cheese varieties also require molds to ripen and develop their characteristic flavours. Asian cuisines rely on a large repertoire of fermented foods. In particular, Aspergillus oryzae and A. sojae, sometimes called koji molds, are employed in many ways. Their hydrolytic enzymes suit them for growth on starch and other carbohydrate-rich substrates. In the koji process, fungal enzymes perform the same function as the malting enzymes used in the beer fermentations of western cultures.

The koji molds release amylases that break down rice starch, which in turn can be fermented to make rice wine. Fermented rice beverages have numerous local variations and names, depending on country and region. Rice wine is called shaoshing in parts of China, sake in Japan, takj or yakju in Korea, as well as by many other names across Asia. The koji molds are also effective in a variety of legume fermentations, of which miso and soy sauce are best known.

Miso is a mixture of soybeans and cereals usually used to flavour soups. Soy sauce is a flavourful, salty liquid sauce made from soybeans that have been fermented by koji molds, yeasts, as well as several halophilic bacteria. Other names for soy sauce include jiangyou (China), makjang and kanjang (Korea), toyo (Philippines) and siiu (Thailand).

Probiotics

Probiotics are living organisms that, when consumed, have beneficial health benefits outside their inherent nutritional effects. There is a growing body of evidence for the role of probiotics in gastrointestinal infections, irritable bowel syndrome and inflammatory bowel disease. Lactobacillus species are used for the production of yogurt, cheese, sauerkraut, pickles, beer, wine, cider, kimchi, chocolate and other fermented foods, as well as animal feeds such as silage. In recent years, much interest has been shown in the use of lactobacilli as probiotic organisms and their potential for disease prevention in humans and animals.Bifidobacteria are considered as important probiotics, and are used in the food industry to relieve and treat many intestinal disorders.

Bifidobacteria exert a range of beneficial health effects, including the regulation of intestinal microbial homeostasis, the inhibition of pathogens and harmful bacteria that colonize and/or infect the gut mucosa, the modulation of local and systemic immune responses, the repression of procarcinogenic enzymatic activities within the microbiota, the production of vitamins, and the bioconversion of a number of dietary compounds into bioactive molecules.

Microbial biopolymers

A variety of biopolymers, such as polysaccharides, polyesters and polyamides, are naturally produced by microorganisms. Several microbially-produced polymers are used in the food industry.

Xanthan

Plant-pathogenic bacteria of the genus Xanthomonas are able to produce the acidic exopolysaccharide xanthan gum. Because of its physical properties, it is widely used as a viscosifer, thickener, emulsifier or stabilizer in the food industry. Xanthan consists of pentasaccharide repeat units composed of D-glucosyl, D-mannosyl, and D-glucuronyl acid residues in a molar ratio of 2:2:1 and variable proportions of O-acetyl and pyruvyl residues.

Alginate

Alginate is the main representative of a family of polysaccharides that neither show branching nor repeating blocks or unit patterns and this property distinguishes it from to other polymers like xanthan or dextran. Alginates can be used as thickening agents. Although listed here under the category 'Microbial polysaccharides', commercial alginates are currently only produced by extraction from brown seaweeds such as Laminaria hyperborea or L. japonica.

Cellulose

Cellulose is a simple polysaccharide, in that it consists only of one type of sugar (glucose), and the units are linearly arranged and linked together by ?-1,4 linkages only. The mechanism of biosynthesis is, however, rather complex, partly because in native celluloses, the chains are organized as highly ordered water-insoluble fibres.

Currently, the key genes involved in cellulose biosynthesis and regulation are known in a number of bacteria, but many details of the biochemistry of its biosynthesis are still not clear. In spite of the enormous abundance of cellulose in plants, bacterial celluloses are being investigated for industrial exploitations.

Poly-γ-glutamic Acid

Poly-γ-glutamic acid (ã-PGA) produced by various strains of Bacillus has potential applications as a thickener in the food industry.

Levan

Levan, a homopolysaccharide composed of D-fructofuranosyl residues joined by 2,6 with multiple branches by 2,1 linkages, has great potential as a functional biopolymer in foods, feeds, cosmetics, and the pharmaceutical and chemical industries. Levan can be used as food or a feed additive with prebiotic and hypocholesterolemic effects.

Exopolysaccharides

Microorganisms synthesize a wide spectrum of multifunctional polysaccharides, including intracellular polysaccharides, structural polysaccharides and extracellular polysaccharides or exopolysaccharides (EPSs). EPSs generally consist of monosaccharides and some noncarbohydrate substituents (such as acetate, pyruvate, succinate, and phosphate). Owing to the wide diversity in composition, they have found multifarious applications in various food and pharmaceutical industries.

Foodborne pathogens

Foodborne pathogens are the leading causes of illness and death in less developed countries, killing approximately 1.8 million people annually. In developed countries, foodborne pathogens are responsible for millions of cases of infectious gastrointestinal diseases each year, costing billions of dollars in medical care and lost productivity.

New foodborne pathogens and foodborne diseases are likely to emerge, driven by factors such as pathogen evolution, changes in agricultural and food manufacturing practices, and changes to the human host status. There are growing concerns that terrorists could use pathogens to contaminate food and water supplies in attempts to incapacitate thousands of people and disrupt economic growth.

Enteric Viruses

Food and waterborne viruses contribute to a substantial number of illnesses throughout the world. Among those most commonly known are hepatitis A virus, rotavirus, astrovirus, enteric adenovirus, hepatitis E virus, and the human caliciviruses consisting of the noroviruses and the Sapporo viruses. This diverse group is transmitted by the fecal-oral route, often by ingestion of contaminated food and water.

Protozoan Parasites

Protozoan parasites associated with food and water can cause illness in humans. Although parasites are more commonly found in developing countries, developed countries have also experienced several foodborne outbreaks. Contaminants may be inadvertently introduced to the foods by inadequate handling practices, either on the farm or during processing of foods. Protozoan parasites can be found worldwide, either infecting wild animals or in water and contaminating crops grown for human consumption. The disease can be much more severe and prolonged in immunocompromissed individuals.

Mycotoxins

Molds produce mycotoxins, which are secondary metabolites that can cause acute or chronic diseases in humans when ingested from contaminated foods. Potential diseases include cancers and tumors in different organs (heart, liver, kidney, nerves), gastrointestinal disturbances, alteration of the immune system, and reproductive problems.

Species of Aspergillus, Fusarium, Penicillium, and Claviceps grow in agricultural commodities or foods and produce the mycotoxins such as aflatoxins, deoxynivalenol, ochratoxin A, fumonisins, ergot alkaloids, T-2 toxin, and zearalenone and other minor mycotoxins such as cyclopiazonic acid and patulin. Mycotoxins occur mainly in cereal grains (barley, maize, rye, wheat), coffee, dairy products, fruits, nuts and spices.

Control of mycotoxins in foods has focused on minimizing mycotoxin production in the field, during storage or destruction once produced. Monitoring foods for mycotoxins is important to manage strategies such as regulations and guidelines, which are used by 77 countries, and for developing exposure assessments essential for accurate risk characterization. Aflatoxins are still recognized as the most important mycotoxins.

They are synthesized by only a few Aspergillus species, of which A. flavus and A. parasiticus are the most problematic. The expression of aflatoxin-related diseases is influenced by factors such as age, nutrition, sex, species and the possibility of concurrent exposure to other toxins. The main target organ in mammals is the liver, so aflatoxicosis is primarily a hepatic disease. Conditions increasing the likelihood of aflatoxicosis in humans include limited availability of food, environmental conditions that favour mold growth on foodstuffs, and lack of regulatory systems for aflatoxin monitoring and control.

Yersinia Enterocolitica

Yersinia enterocolitica includes pathogens and environmental strains that are ubiquitous in terrestrial and fresh water ecosystems. Evidence from large outbreaks of yersiniosis and from epidemiological studies of sporadic cases has shown that Y. enterocolitica is a foodborne pathogen. Pork is often implicated as the source of infection. The pig is the only animal consumed by man that regularly harbors pathogenic Y. enterocolitica.

An important property of the bacterium is its ability to multiply at temperatures near 0°C, and therefore in many chilled foods. The pathogenic serovars (mainly O:3, O:5, 27, O:8 and O:9) show different geographical distribution. However, the appearance of strains of serovars O:3 and O:9 in Europe, Japan in the 1970s, and in North America by the end of the 1980s, is an example of a global pandemic. There is a possible risk of reactive arthritis following infection with Y. enterocolitica.

Vibrio

Vibrio species are prevalent in estuarine and marine environments, and seven species can cause foodborne infections associated with seafood. Vibrio cholerae O1 and O139 serovtypes produce cholera toxin and are agents of cholera. However, fecal-oral route infections in the terrestrial environment are responsible for epidemic cholera. V. cholerae non-O1/O139 strains may cause gastroenteritis through production of known toxins or unknown mechanism.

Vibrio parahaemolytitcus strains capable of producing thermostable direct hemolysin (TDH) and/or TDH-related hemolysin are most important causes of gastroenteritis associated with seafood consumption. Vibrio vulnificus is responsible for seafoodborne primary septicemia, and its infectivity depends primarily on the risk factors of the host.

V. vulnificus infection has the highest case fatality rate (50%) of any foodborne pathogen. Four other species (V. mimicus, V. hollisae, V. fluvialis, and V. furnissii) can cause gastroenteritis. Some strains of these species produce known toxins, but the pathogenic mechanism is largely not understood. The ecology of and detection and control methods for all seafoodborne Vibrio pathogens are essentially similar.

Staphylococcus Aureus

Staphylococcus aureus is a common cause of bacterial foodborne disease worldwide. Symptoms include vomiting and diarrhea that occur shortly after ingestion of S. aureus toxin-contaminated food. The symptoms arise from ingestion of preformed enterotoxin, which accounts for the short incubation time. Staphylococcal enterotoxins are superantigens and, as such, have adverse effects on the immune system.

The enterotoxin genes are accessory genetic elements in S. aureus, meaning not all strains of this organism are enterotoxin-producing. The enterotoxin genes are found on prophages, plasmids, and pathogenicity islands in different strains of S. aureus. Expression of the enterotoxin genes is often under the control of global virulence gene regulatory systems.

Campylobacter

Campylobacter spp., primarily C. jejuni subsp. jejuni is one of the major causes of bacterial gastroenteritis in the U.S. and worldwide. Campylobacter infection is primarily a foodborne illness, usually without complications; however, serious sequelae, such as Guillain-Barre Syndrome, occur in a small subset of infected patients. Detection of C. jejuni in clinical samples is readily accomplished by culture and nonculture methods.

Listeria Monocytogenes

Listeria monocytogenes is Gram-positive foodborne bacterial pathogen and the causative agent of human listeriosis. Listeria infections are acquired primarily through the consumption of contaminated foods, including soft cheese, raw milk, deli salads, and ready-to-eat foods such as luncheon meats and frankfurters. Although L. monocytogenes infection is usually limited to individuals that are immunocompromised, the high mortality rate associated with human listeriosis makes it the leading cause of death among foodborne bacterial pathogens. As a result, tremendous effort has been made to develop methods for the isolation, detection and control of L. monocytogenes in foods.

Salmonella

Salmonella serotypes continue to be a prominent threat to food safety worldwide. Infections are commonly acquired by animal to human transmission though consumption of undercooked food products derived from livestock or domestic fowl. The second half of the 20th century saw the emergence of Salmonella serotypes that became associated with new food sources (*i.e.* chicken eggs) and the emergence of Salmonella serotypes with resistance against multiple antibiotics.

Shigella

Shigella species are members of the family Enterobacteriaceae and are Gram negative, nonmotile rods. Four subgroups exist based on O-antigen structure and biochemical properties: S. dysenteriae (subgroup A), S. flexneri (subgroup B), S. boydii (subgroup C) and S. sonnei (subgroup D). Symptoms include mild to severe diarrhea with or without blood, fever, tenesmus and abdominal pain. Further complications of the disease may be seizures, toxic megacolon, reactive arthritis and hemolytic uremic syndrome. Transmission of the pathogen is by the fecal-oral route, commonly through food and water.

The infectious dose ranges from 10-100 organisms. Shigella spp. have a sophisticated pathogenic mechanism to invade colonic epithelial cells of the host, man and higher primates, and the ability to multiply intracellularly and spread from cell to adjacent cell via actin polymerization. Shigella spp. are one of the leading causes of bacterial foodborne illnesses and can spread quickly within a population.

Escherichia Coli

More information is available concerning Escherichia coli than any other organism, thus making E. coli the most thoroughly studied species in the microbial world. For many years, E. coli was considered a commensal of human and animal intestinal tracts with low virulence potential. It is now known that many strains of E. coli act as pathogens, inducing serious gastrointestinal diseases and even death in humans.

There are six major categories of E. coli strains that cause enteric diseases in humans, including the:

- Enterohemorrhagic E. coli, which cause hemorrhagic colitis and hemolytic uremic syndrome,
- Enterotoxigenic E. coli, which induce traveller's diarrhea,
- Enteropathogenic E. coli, which cause a persistent diarrhea in children living in developing countries,
- Enteroaggregative E. coli, which provokes diarrhea in children,
- Enteroinvasive E. coli that are biochemically and genetically related to Shigella species and can induce diarrhea,
- Diffusely adherent E. coli, which cause diarrhea and are distinguished by a characteristic type of adherence to mammalian cells.

Clostridium Botulinum and Clostridium Perfringens

Clostridium botulinum produces extremely potent neurotoxins that result in the severe neuroparalytic disease, botulism. The enterotoxin produced by C. perfringens during sporulation of vegetative cells in the host intestine results in debilitating acute diarrhea and abdominal pain.

Sales of refrigerated, processed foods of extended durability including sous-vide foods, chilled ready-to-eat meals, and cook-chill foods have increased over recent years. Anaerobic spore-formers have been identified as the primary microbiological concerns in these foods. Heightened awareness over intentional food source tampering with botulinum neurotoxin has arisen with respect to genes encoding the toxins that are capable of transfer to nontoxigenic clostridia.

Bacillus Cereus

The Bacillus cereus group comprises six members: B. anthracis, B. cereus, B. mycoides, B. pseudomycoides, B. thuringiensis and B. weihenstephanensis. These species are closely related and should be placed within one species, except for B. anthracis that possesses specific large virulence plasmids. B. cereus is a normal soil inhabitant, and is frequently isolated from a variety of foods, including vegetables, dairy products and meat. It causes a vomiting or diarrhea illness that is becoming increasingly important in the industrialized world. Some patients may experience both types of illness simultaneously. The diarrheal type of illness is most prevalent in the western hemisphere, whereas the emetic type is most prevalent in Japan. Desserts, meat dishes, and dairy products are the foods most frequently associated with diarrheal illness, whereas rice and pasta are the most common vehicles of emetic illness.

The emetic toxin (cereulide) has been isolated and characterized; it is a small ring peptide synthesised nonribosomally by a peptide synthetase.

Three types of B. cereus enterotoxins involved in foodborne outbreaks have been identified. Two of these enterotoxins are three-component proteins and are related, while the last is a one-component protein (CytK). Deaths have been recorded both by strains that produce the emetic toxin and by a strain producing only CytK.

Some strains of the B. cereus group are able to grow at refrigeration temperatures. These variants raise concern about the safety of cooked, refrigerated foods with an extended shelf life. B. cereus spores adhere to many surfaces and survive normal washing and disinfection (except for hypochlorite and UVC) procedures. B. cereus food borne illness is likely under-reported because of its relatively mild symptoms, which are of short duration.

Food authenticity

It is important to be able to detect microorganisms in food, in particular pathogenic microorganisms or genetically modified microorganisms. Real-time PCR is an accepted analytical tool within the food industry. Its principal role has been one of assisting the legislative authorities, major manufacturers and retailers to confirm the authenticity of foods. The most obvious role is the detection of genetically modified organisms, but real-time PCR makes a significant contribution to other areas of the food industry, including food safety.

THE SITE FOR CROP PRODUCTION

Isolation and Rotation

The considerations here are similar in principle to those stated above for the nursery. The site should be at least 200 yards (approximately 200 meters) from any existing or planned related crop such as tomato, potato, tobacco or other pepper crop. This is to minimize the risk of insect pests and their transmission of virus diseases.

Crop rotation should be practiced to ensure that the selected field has not grown a pepper, potato, tomato or tobacco crop for at least two years. A plot that has been fallow for two years would be ideal provided it has all the other requirements listed here. In some pepper production regions farmers are advised to have a four year rotation for crops in the pepper family.

Soil: Although peppers will do reasonably well on most soils there are some important points to appreciate. The top soil should be deep and well drained; a soil that becomes too wet or waterlogged is likely to cause leaf drop.

The crop needs a good nutrient supply which can be helped by the addition of organic materials such as manures, compost and decomposed mulches. Materials such as old crop straw or residues will use some of the available nitrogen during their break down unless they are rotted or decomposed first. The addition of bulky organic materials will improve the soil's moisture holding capacity, especially in the long term.

The soil pH should be between 6.0-7.0, ideally 6.0-6.5. The pH of the field should be tested from typical soil samples and a liming material added if required during preparation. The soil's nutrient levels should also be checked from the soil samples. The optimum N:P:K ratio is 1:1:1.5, an excess of nitrogen will delay the start of flowering and therefore result in later fruit production. In those areas with a frequent high rainfall which can result in nitrogen being rapidly leached (washed out of the soil), approximately half of the basic nitrogen requirement can be applied later as top dressings, but not after the start of flowering except in long term crops.

Preparing the site: The land selected for the pepper crop should be ploughed in advance of surface preparation. This increase in surface area will allow some natural weathering and drying out to take place which is especially important in poorly drained soils. It will also allow birds to feed from some of the soil-borne pests. If the field has poor drainage it is preferable to grow the crop on mounds or prepared ridges.

The ridges should be prepared 36-40 inches (approximately 90-100 cm) apart and 6-8 inches (15-20 cm) high. If possible the ridges should slope down the field to improve surface runoff and drainage; this is especially important in areas which are likely to have heavy rain storms with surface runoff. However, if the slope is too great there is a greater risk of the soil being washed away when this occurs. Some farmers allow grass to grow in the gullies, trenches or drains so as to minimise soil erosion but this cannot be done if a contact weed killer is going to be applied between the pepper plants during crop production.

3

Agriculture and Food Security

INTRODUCTION

A number of countries are now undertaking ambitious efforts to achieve the New Vision through a transformation of their agriculture sectors, engaging in public-private collaboration and market-based approaches.

The report outlines six elements that are essential stepping stones toward success in a large-scale transformation effort. These are illustrated by examples from national initiatives underway in Asia, Africa and Latin America.

RISK MITIGATION IN AGRIBUSINESS

India has 20 regional exchanges dealing in one/few commodities. Recently the Government allowed multi commodity exchanges to operate in the Indian market. The Government has permitted trading in 120 commodity and commodity derivatives in the futures market. However, the commodity exchanges face several bottlenecks which include:

- Regional Exchanges
- Small number of participants
- Dominance of brokers; Cartelisation
- Lack of trust in operations (especially clearing) of futures exchanges
- Low incidence of transactions with physical settlement
- Commercial viability of exchanges under question as each exchange handles one/few commodities-lack of critical transaction volumes.

Others

- Forward Markets Commission is not a well-funded and strong body with powers for regulation

- Government policies impacting real price discovery
- Minimum support prices, which impact the price level, and render Indian prices out of line with international prices
- Lack of clear guidelines on proper tax recognition of hedging
- High cost of risk management due to stamp duties in some states
- Government ban on use of futures by banks
- Options are not allowed to be traded
- Warehouse Receipts are not negotiable and freely transferable instruments
- Lack of standardization of commodities

Some of the solution themes identified which can assist in development of a robust futures market are as follows:

Introduction of Options Trading

Equip the FMC with penal powers for arbitration

Removal of Regulatory Bottlenecks

- Amendment required in FCRA 1952 for the following
- Changing the definition of commodity to be an exclusive definition (for instance "anything not a security" as defined in the USA)
- Bifurcation of the commodity into (i) goods & (ii) non-goods-all existing commodities will be goods, while non-goods will include indices, weather etc. non-goods which will be allowed to be cash settled.

Futures Markets

- Mandi prices should be provided on a real time basis to all futures exchanges
- All exchanges should adopt on-line trading to allow access even at remote trading centres and thus ensure better participation and price discovery

Improve Physical Settlement Procedures

- Standardization of commodities-Institute mechanism for grading of various traded commodities
- Strengthen Warehouse Receipts System: In addition to the measures listed earlier, in the context of futures trading, the FMC needs to make it mandatory for all exchanges to either certify warehouse operators or set up a professional collateral management body

Remove Taxation Hurdles

- Allow set-off of commodity futures trading against business profits/ losses to enable farmers to participate in the exchanges
- Income tax rules mandate that an entity engaged in hedging should have a firm contract in the underlying commodity or have possession of the goods against which hedging is being undertaken, in the absence of which the entity is liable to pay capital gains tax (30%).The rules should be amended such that the trader can submit the proof of the actual possession of the commodity at a future date, when the derivatives transaction is being concluded.

Clarity on Policy Framework

Frequent and sudden duty changes on commodities hampers trading. The policy parameters for changes in duty structure in the medium term, should be well-established to enable players to operate effectively.

HOUSEHOLD FOOD SECURITY

The nutritional status of each member of the household depends on several conditions being met: the food available to the household must be shared according to individual needs; the food must be of sufficient variety, quality and safety; and each family member must have good health status in order to benefit from the food consumed. Agriculture development and food production are central to national food security. In addition, there are several other factors that determine the food security at household level. These include food prices, food availability, employment and purchasing power, government subsidies and the food rationing system.

Food insecurity continues to threaten large proportions of households in low income countries. It is common among the absolute poor in middle income countries, and even in some rich countries. The problem is widespread, and is not confined to any one sector or group of nations. Even when hunger is avoided, families suffer from its threat. The entire society benefits when people feel their access to food is secure. An operational definition of household food security is proposed as follows. A household is food secure when it has access to the food needed for a healthy life for all its members (adequate in terms of quality, quantity, safety and culturally acceptable), and when it is not at undue risk of losing such access.

Food insecurity, as a household-level issue, can be addressed by a wide range of alternative policies and combinations of policies and programmes. Policies for food security should aim at attaining required food consumption levels and reducing the risk of the poor losing access to food. Access to food and purchasing power are central, and both transitory (e.g. seasonal) and chronic food insecurity problems are of concern. Adequate global and national level food supply remain necessary but insufficient conditions for household food security. High levels of food self sufficiency in low income countries have no necessary relationship to their households' food security, which has to be addressed by specific policies. Households should be viewed in the context of their community, and not in isolation. Many of the problems considered below have an important community and local government dimension and cannot be addressed by the central government alone.

The nature and scale of the food security problem differs a great deal among and within countries, and also between urban and rural areas. Wise policy needs to take account of this, and therefore has to be country-and region-specific and problem-oriented. Food security cannot be achieved free of charge in terms of fiscal resources. Public capabilities for problem identification and policy design and implementation are required to help to find ways to eliminate the unacceptable human misery caused by food insecurity, or by extreme efforts that households may take to avert it. In addition to humanitarian considerations, food secure households are a precondition for a modernizing and healthy society whose members concern themselves with investment in a productive future (e.g. education) rather than scrambling for adequate food today. Governments have an obligation to enable families and communities to achieve long-term food security and to provide a safety net to prevent destitution.

Food is such a high priority for poor households that many may be tenuously "secure", but at great sacrifice-for example spending almost all their money or time on securing food. Thus not only must current food security itself be tackled, but also both the vulnerability and the disadvantages from enforced concentration on acquiring food, to the detriment of other needs like education or housing. Moreover increases in income even among the lowest income groups do not necessarily go entirely to increasing food energy intakes, but also towards better quality in terms of a more palatable and diversified diet. This represents an important aspiration not captured by dietary energy intakes alone, and is another objective of improved food security.

Measurement of Household Food Security

While the basic concept of household food security is clear, and an ideal measure of it is easy to describe, it is surprisingly difficult to gauge it in practice. This difficulty does not mean waiting for years of academic research. It does suggest that operational research and evaluation should be built into food security activities, so that confidence in the precision of actions will grow as the decade progresses. An ideal measure of household food security includes the measurement of household food availability and average household food consumption levels over a period of time, in relation to need. For various reasons, this is all but impossible to achieve at a reasonable cost in a reasonable time period: there are problems with measuring both availability and consumption, and need itself. It is sufficiently difficult that it is best regarded as an ideal rather than practical measure. The proportion of available resources required for achieving food security may also be assessed; for example, households with adequate food security but spending almost all their income on food should clearly be distinguished from those only needing to spend a moderate proportion on food. This proportion is indicative of the stress on households' well-being, and reflects on their capacity to cope and indeed survive.

There are a number of other variables that might help to indicate trends, or serve as proxies for, food security. The best general indicator is probably real income, although still hard to assess; more fundamental measures, such as landlessness, should be included. Research is needed to see which groups of indirect indicators are best used under which conditions. In general, more weight has to be placed on indirect indicators where local government is weak and participation in the policy process is low; and also where investigative journalism is suppressed. Examples of potential indicators are changes in food production by region, changes in price ratios (e.g. crop/livestock or crop/wage price ratios), migration, assets; priority data from household surveys conducted for this specific purpose include; food consumption, the perceived risk of food insecurity, use of famine foods, and anthropometric measurements. In relation to causality, none of these indicators are reliable on their own. All should be used in conjunction with other information. Emphasis should be on changes from normal levels, as many indicators will change for reasons unrelated to food security. For example, migration may rise in response to urban job opportunities, and weight-for-length indices may drop due to a rise in infectious disease. In general, careful analysis is needed before inferring changes in food security. However, there is often adequate data available or easily gathered to allow judgements about food security-lack of precise data is no excuse for inaction.

Poverty

The poorest groups in India tend to be asset-less or nearly asset-less. Landlessness is increasing, both absolutely and relatively (from 9.6 per cent of rural population in 1971 to 11.3 per cent in 1982: GOI 1988). The poor derive income from wage work in which they may be simultaneously underemployed, underpaid and forced into unwaged work. The path of diversification of the rural economy varies from region to region and is often insufficient to absorb surplus labour from agriculture. In addition, there is still a particular institutionalized poverty amongst scheduled castes and tribes, and in regions where such people are concentrated the rates of poverty reduction are also very low. Within poor households, and more extensively throughout society in northern India, women and children are particularly disadvantaged and female-headed households are amongst the very poorest.

Actual measurement of poverty in the Indian context has come under debate in recent years. There are two conventional approaches: one based on average annual household income (in relation to an income poverty line defining a predominantly caloric threshold) and the other on calorie intake. Although income is widely used to measure economic welfare, it has serious drawbacks, a major one being that it may have substantial fluctuations which are averaged out in the long run. Consumption expenditure may be a better indicator of the actual economic position of a household (Kakwani and Subbarao 1990). The National Sample Surveys provide reasonably comparable time series data on the levels and distribution of household consumption expenditure.

The Indian Planning Commission estimates the population below a national poverty line, defined in 1981 in the Sixth Five Year Plan as being the mid-point of the monthly expenditure class having a daily per capita calorie intake of 2,400 kcals. in rural areas and 2100 kcals in urban areas. The incidence of poverty (so defined) declined steadily from 51.5% in 1972-73 to 29.2% in 1987-88-a fall of over 22%. Estimates by economists, however, suggest a much more modest decline of a little above 10% from 56.3 to 45.9%. The Planning Commission reports a decline in total numbers of the poor from 291.5 million to 232.4 million while other estimates suggest a substantial increase from 308 million to 361.2 million. While both estimates suggest a decline in the *prevalence* of poverty, the quantum of the decline implied by the Planning Commission is far in excess of those from alternate and possibly more realistic estimates.

The reasons for the disparity between the two estimates lies in the different definition of the poverty line and "adjustments/corrections" applied to the NSS data. The Planning Commission draws the line at Rs 49.08 and Rs 56.64 for rural and urban areas respectively at 1973-74 prices. Further, NSS food consumption data are scaled upwards by the Planning Commission to adjust for apparent underestimations, as estimated by comparison of NSS data with those from the Central Statistical Organization. This exercise results in "pulling up" a large number of persons who would otherwise be below the poverty line-a procedure disputed by many economists. Alternate estimations of poverty use state-specific cost of living indices for the middle range of the population "falling within the range of monthly per capita total expenditure which encloses the state-specific poverty norm in the base year" (Suryanarayana, 1991).

The state-specific criteria are used to yield state-specific poverty lines and head count ratios from state-specific NSS distributions. These considerations are important in assessing the relative success of poverty alleviation programmes in the last two decades.

Furthermore, as argued by *Suryanarana* (1991), if crossing the poverty line is used as the sole estimate of the success of poverty alleviation programmes, administrators may be inclined to focus only on the segment of population just below the poverty line. There is therefore a need to use distributionally-sensitive poverty indices in assessing the effectiveness of poverty alleviation programmes. Some attempts at using a similar approach are reflected in Subbarao's (1989) assessment of the changes in nutrient intakes among the poor and the "ultra poor". Another way of investigating change is to focus instead on the extent to which the rich and the poor in rural India have been able to increase their consumption over time. The following pattern emerges from such an exercise, reported in the national newspaper, *The Hindu* (Ranade December 18, 1991):

Ranade notes that the quadrupling of the rate of growth in consumption of the rural poor was near-synchronous with the large increase in 'targeted' poverty alleviation programmes in the early 1980s.

Any 'trickle down' effect of overall growth during this period was unlikely, given the deceleration in the rate of growth of the rural rich. This also concurs with evidence provided by Kakwani and Subbarao (1990) which suggests that the improvement between 1977 and 1983 was not associated with increased foodgrain production, but rather a decrease in within-state inequality along with the likely beneficial effects of several large-scale poverty-alleviation programmes. The improvement observed later, in the mid-1980s, on the other hand was likely to have been much more driven by economic growth and favourable relative food price changes. Kakwani and Subbarao (1990) went on to show that this reduction in inequality was more pronounced for the ultra-poor than the poor. Ranade however questions whether the type of poverty alleviation programmes implemented in the early 1980s could be sustained given the poor state of the economy in the early 1990s. There are also inter-state variations in the effectiveness of these programmes.

Kakwani and Subbarao (1990) also found that between 1972 and 1983 inequality at the all-India level, is driven largely by within-state, not between-state, inequality, although this appears slowly to be changing-between-state inequality is rising. In 1983, the states could be ranked thus, in order of declining poverty (using the head-count method): Bihar, West Bengal, Tamil Nadu, Orissa, Madhya Pradesh, Uttar Pradesh, Maharasthra, Assam, Kamataka, Kerala, Rajasthan, Andhra Pradesh, Gujarat, Haryana, Punjab. In fact, this ranking is almost identical to that of 'ultra poverty'. Looking at the map of India, poverty is apparently concentrated in the north east of the country (Bihar, West Bengal) and, apart from Tamil Nadu in the south, stretches out to cover the neighbouring states of Orissa, Madhya Pradesh and Uttar Pradesh.

Food prices and expenditure

The trends in food price indices, based on data from the Economic Survey of India (GOI 1991). While absolute food prices have risen, their levels relative to the consumer price index have fluctuated connsiderably during the 1980s, with a slight overall declining trend (the ratio of food to consumer price indices is an indicator of a consumer's propensity to buy food as opposed to non-food articles). A downwards trend in relative food prices is likely to have a disproportionately beneficial effect on the poorest groups, with the highest price elasticities of foodgrain demand.

Mean elasticities are estimated (Sarma and Gandhi 1990) to be 0.48 for rural, 0.23 for urban, and 0.42 for national, but elasticities differ sharply by quartile, being close to 1.0 for the bottom quartiles and about 0.1 for the top quartiles.

Two thirds of the total expenditure of rural households is spent on food as compared to slightly more than half for urban households (Sarekshama 1989). Between 1972 and 1987, these proportions changed with relatively more money being spent on non-food items in both rural and urban areas; the main change occurring between 1972 and 1978.

National Sample Survey consumption data indicate a small improvement in the distribution of total expenditure between 1977/78 and 1983 in both the rural and urban areas, and between the rural and urban areas. For the poor population, the data indicate that foodgrain consumption of the rural bottom quartile rose somewhat between 1970/71 and 1983, but that of the urban bottom quartile was virtually stagnant. This is associated with a small increase in the real total; expenditure of the rural bottom quartile (combined with high elasticities) but near stagnation in that of the urban bottom quartile.

FOOD SHORTAGES IN NORTHEAST INDIA

The world economy has seen significant growth in the last decade through increases in productivity, product quality, and export base diversification. These advances were mainly driven by the growth of traditional agricultural and industrial sectors. The application of agriculture biotechnology offers the opportunity to alleviate marginal and subsistence farmers to get out of their dependency on government assistance and to increase production. The introduction of new crop varieties with insect and herbicide resistant genes is quickly changing the landscape of agriculture in the US and many countries. Major pests can be eliminated with little or no spray of pesticides, as well as nasty weeds killed with the spray of herbicide in genetically engineered crop varieties. Genetically engineered crop varieties are environmental friendly with reduced chemical use as an added benefit.

Given the need to feed more people on the same land area while using less water and nutrients, the application of biotechnology to improve crop plants is fast becoming the only viable option for the teeming millions. Let me give a direct quote from Dr. Norman Borlaug, Noble Peace laureate, ?the world need to double food production by 2050 if hunger were to be banished from the world and the ongoing 'gene revolution' can definitely play a part in this. You cannot build peace on empty stomachs. Only 8 per cent of countries with lower levels of hunger are mired in conflict". I might add, insurgency, the dominant menace of the region is eating up the resources and turning many vibrant and intelligent youths into dangerous thugs because there is not enough opportunities to go around. Many insurgency groups have sprang up as a result of hunger and a lack of opportunity rather than a genuine desire to leave the society in better shape than the one they currently inherit.

Pied Pipers In North East-India

However, the impact of famine on the people are generally left out in such discourses. To fill up such academic deficiency, Sajal Nag brought out a serious study on the history and politics around the bamboo famine in the context of Mizoram from the 1880s to the 1960s. Among India's north east states, Mizoram occupies a unique position mainly due to the periodic bamboo flowerings which lead to devastating famines in the region. This eventually led to the Famine Relief Front turned into political party called Mizo National Front uprising against the Indian nation in the mid 1960s. Mizo history from the pre-colonial migration to the encounter with colonialism and their interaction with the European Christian Missionaries and the growth of ethno-nationalism in post independent period were strongly determined by the cyclical bamboo famine.

The various famine relief efforts which were made during colonialism by the state were all part of colonial agenda of trying to win over the hearts of the tribal people. In contrast to the colonial policy, the post colonial state ignored the famine in the hills. As a consequence, the socio-political upheaval started in 1966 which lasted for 20 years. Hence, history has provided a great deal of lesson that government interventions become important elements in coping with the famine stress. Famine was not new to the British, particularly since the Bengal famine. In fact, they were responsible for most of the famines that had occurred in India. (p.63) However, bamboo famine in Mizoram (Lushai Hills) has triggered their curiosity as it was new to them.

Bamboo flowering causes two types of famine locally called "mautam tam" and "thingtam tam"-their occurrence usually based on chronological sequences. In local language, Mautam or thingtam literally means bamboos withering or dying out. Tam meaning famine, the famine caused by bamboo is therefore called mautam tam and thingtam tam. According to folk tradition, the two species of bamboo (thing or bambusa longispiculata and mautak or melocanna bambbasoides) that grow in abundance in the hills flower in a cycle of 30 years and 50 years respectively. The bamboo flower soon produces fruits which are usually brown and green in colour. The seeds contain rich protein and are the favourite food of the local rats or Sazu. This is followed by a sudden explosion of rat population in the bamboo hills. The bamboos soon die, leaving the rats without abundant food from the bamboo. The hungry rats soon ravage the entire paddy field and stored grain in a matter of one night ultimately led to famine. The famine caused by bamboo flowering was recorded for the first time in 1862 and the next bamboo famine which devastated the entire Mizo chiefdoms was recorded in 1881-1882 (Military Report on the Chin-Lushai Country, 1881). It was estimated that 15,000 Mizos perished. (p.62) Many people fled to the neighbouring plain areas of Cachar and Manipur. Nag argued that the Mizo tribe who were dependent on jhum agriculture and hunting were helpless and entirely dependent on the external intervention especially in the time of famine. At the same time, colonial government needed to generate knowledge on "savage predators" who continuously attacked their commercial activities in plain areas.

Hence, the year 1881 ushered a new milestone in the Anglo-Mizo relationship as bamboo famine opened a new chapter for the production of colonial knowledge on the Mizo tribe. During the time of famine, not only did the Mizo chiefs agree to cease their hostilities but they also allowed the colonials to enter their hill territory to inquire the condition of the famine. Such inquiry enabled the British to gain valuable information on the Mizo chiefdoms. Soon after the colonial government acquired some knowledge, the first series of "punitive expeditions" were launched in 1840 which continued up to the end of of the 19th century. However, the Mizo Chiefs were not easily subdued despite the repeated efforts on the part of the British government from outright war to economic blockade, from burning stored food and standing crop to taking prisoners. Then the British subsequently transformed their punitive expeditions to wars of conquest as "the tribal raids would not cease unless they were totally conquered and placed under 'civilized' administration of the British and culturally transformed through the quietening influence of Christianity".

WHAT IS GMO?

The acronym GMO stands for ?genetically modified organism, and refers to any plant or animal species that has had a gene or genes from a different species transferred into its genetic material via accepted genetic engineering techniques. The process of introducing genes into a new species and getting them to function is known as transformation. Genetically modified organisms have a foreign gene inserted into them that creates one or more new traits for that organism. The genetic engineering employs an array of methodology such as gene discovery, transformation, molecular characterization of the insert, and gene expression study. In a real sense, genetically modified is not a new phenomenon, but has been practiced by plant breeders for years through breeding and selection with the objective of producing more productive, more resistant offspring, or in producing better or different quality of product than the existing variety. The process to produce new varieties following conventional breeding methodology takes approximately 6-7 years compared to 2-4 years via genetic engineering. A single gene with a known function can be moved from one organism to another without transferring additional unrelated genes, as happens in the process in conventional breeding method. Above all, what marks GMO out from the products of conventional breeding was that the new techniques help us to cross taxonomic lines (across unrelated species), which means I can put a B.t. gene from a soil bacteria over to cotton.

HOW SAFE ARE GMO CROPS?

The majority of scientists have consensus that there is nothing inherently risky about genetically engineered crops. Splicing genes from one organism into another has been used for years. One example is the production of an enzyme required in processing cheese. No one calls that dangerous. Recently introduced genetically engineered rice (golden rice) that could potentially eliminate vitamin A deficiencies has been met with widespread acceptance. Before a variety with a new trait is approved for cultivation, it has to go through rigorous government agency approval process. For example, in the US, three agencies: the Food and Drug Administration (FDA), United States Department of Agriculture (USDA), and Environmental Protection Agency (EPA) must approve the new gene(s) before the traits are released. The process is very involved and takes up to 2 years for final approval. The agencies ask the following questions: is it safe to consume (FDA); is it safe to grow (USDA); and is it safe for the environment (EPA).

THE BENEFITS OF GMO

The best way to demonstrate the benefits of GMO is through examples.

(1) The virus that causes severe ring spot in papaya. The disease reduces papaya production and kills the trees in Asia, in parts of Latin America, and in Africa.
(2) Leaf curls disease on white potatoes.
(3) The leaf yellowing in sweet potatoes throughout east and central Africa.
(4) The stunting and yellowing in rice throughout central Asia.

Each of these examples is caused by different strains of virus. Each of these important diseases can be controlled through biotechnologies that increase the resistance of the plants to the viruses. The production of cotton in India, Pakistan, Egypt and other countries where the boll worm, boll weevil and other insect pests have in the past reduced yields, have with the application of biotechnology turned around and increased.

The common insect resistant trait gene used in the market place today is called B.t. gene. When smallholder farmers in China and South Africa grew native cotton varieties that contained the B.t. gene for insect resistance as a result of biotechnology, the farmers realized between $150 and $200 per hectare in increased profits. It is estimated that more than a million farmers (combined) in these two countries have benefited from insect resistant varieties of cotton. The increased profit came because of the reduced application of insecticides to control the pests and increased lint production. Perhaps the most striking examples of how biotechnology can improve human nutrition are found in the many varieties of rice and canola that have been improved by biotechnology. Biotechnology has been used in rice and canola to increase the amounts of beta-carotene, the precursor of Vitamin A, which is in short supply in diets in many parts of the world. The hope is that consumption of foods rich in beta-carotene will alleviate the chronic Vitamin A deficiencies in the diets of many of the poor in Asia and Africa. Other research is underway to increase the levels of other vitamins, to improve the amount of proteins in crops like potatoes and cassava. Researchers are also developing foods that can deliver certain types of therapeutic substances, such as vaccines, which stimulate the body's immune system against certain endemic diseases.

BIOTECHNOLOGY CENTER IN NE INDIA

The NE region is rich in diversity-both culturally and agriculturally. With agriculture being the predominant industry of the region, any positive changes focused on here is welcome news for 70% of the population. Since land holding is limited, any increase in agriculture product must come from either changing the cultural practices or integration of agricultural biotechnology. The later is attractive and promising for the future since there is a limit to what cultural practices will add to the overall production. For example, the Green Revolution in the early 1960s with the introduction of dwarf varieties of wheat and rice depended on cultural practices to make the changes, but production leveled off after agricultural inputs where enacted.

Establishing a biotechnology center is imperative if the region is to improve the agriculture to meet the future food demands of the region. Crop improvement through genetic engineering must be a priority if the region wants to stay competitive and to best utilize natural resources. A number of states have already established their own biotechnology center as is in the state of Andhra Pradesh. The region already has a number of universities that can provide the technical know-how and skills needed in the hundreds of competent faculty and scientists they employ. Additionally, there are several Indian Council of Agricultural Research centers in the region, which may have already thought through the process of establishing a biotechnology center. For a biotechnology center to stay competitive, the center must reflect the following criteria:

(1) as a government (central and regional) sponsor with a non-political entity,
(2) governed by a board of directors (scientist should make up a majority of the board),
(3) registered as a non-profit organization,
(4) development process driven by product delivery rather than consumption driven,
(5) seek its own funding through government and foundation, both national and international,
(6) competitive funding scheme should ensure the integration of biotechnology with classical breeding and other associated technical disciplines,

(7) located close in proximity to a major airport,
(8) close to a major university or college,
(9) have the goal of becoming a self-supporting entity,
(10) have the objective of addressing regional issues,
(11) employee selection criteria strictly merit, and
(12) the seven sister states each contribute and play a role.

ECONOMIC IMPACTS

The economic impact on the region is enormous-from food import dependency to a player in India and the world food market. The total area under GM crops globally has risen by 20 per cent in the last year alone. In India, where GM use was allowed only three years ago, about 1.34 million hectares have come under B.t. cotton. A cotton farmer in India can increase his bottom line even though he pays higher seed premium in order to grow genetically engineered cotton. The market outlook for GM products in India is excellent since food shortage and poverty are still alive in India.

In conclusion, I believe biotechnology can bring unprecedented economic impact in the region, which in turn will change the attitude of the people. I believe if a product and market driven economy takes a hold in the region, the young people will trade their guns for the newfound economic power they so desperately seek to achieve. Rather than stealing at gunpoint, they will work hard exploiting the new economic reality within their reach.

INDUSTRIAL BIOTECHNOLOGY

At present, a third wave of biotechnology-industrial biotechnology-is strongly developing. Industrial biotechnology (also referred to as white biotechnology) uses biological systems for the production of useful chemical entities. This technology is mainly based on biocatalysis and fermentation technology in combination with recent breakthroughs in the forefront of molecular genetics and metabolic engineering. This new technology has developed into a main contributor to the so-called green chemistry, in which renewable resources such as sugars or vegetable oils are converted into a wide variety of chemical substances such as fine and bulk chemicals, pharmaceuticals, bio-colorants, solvents, bio-plastics, vitamins, food additives, bio-pesticides and bio-fuels such as bio-ethanol and bio-diesel.

The application of industrial biotechnology offers significant ecological advantages. Agricultural crops are used starting raw materials, instead of using fossil resources such as crude oil and gas. This technology consequently has a beneficial effect on greenhouse gas emissions and at the same time supports the agricultural sector producing these raw materials. Industrial biotechnology frequently shows significant performance benefits compared to conventional chemical technology.

Strategic Actions:

- Focus in industrial biotechnology will be on reducing chemical and toxic load in our effluent streams, developing non-fossil fuels that are eco-friendly and developing green technologies in Industrial processing.
- Encourage public-private partnership to promote investment in this sector.
- Promotion of industrial biotechnology in strategic areas of manufacturing and developing green technologies.

DIAGNOSTICS FOR EMERGING MEDICAL PARADIGM

There is potential to generate a new repertoire of tools for screening people for risk of disease, for early detection of infections and chronic diseases and for predicting outcome. In certain circumstances, single tests are requires to detect multiple pathogens or biochemical abnormalities. To be widely useful, diagnostics need to be real time and low cost. Advances in biosensors and gene amplification are in the offing to enable real time medicine. Immuno proteomics has the potential to reveal multiple targets for development of diagnostics for diseases for which existing tools are unsatisfactory. For chronic diseases, a shift from treating disease on an individual basis is visualized by genetic assessment of likelihood of benefit from a therapeutic intervention, the so-called personalized medicine. It is seen that most drugs work in only a proportion of patients, targeting therapy to the right sub-group will not only make therapy more efficacious but also make evaluation of newer products cheaper.

Pharmacogenomics is a rapidly growing segment that provides a wealth of information pertaining to defective or missing genes, which call for differentiated medicine-a new avenue for drug research. This emerging discipline combines both infotech and biotech skills in augmenting high-speed data mining of both genotypic and phenotypic information with a view to evolving new forms of medical diagnostics and therapies. Gene regulation and other bio-algorithms will form the core of a new wave of diagnostics that are now being referred to as 'theranostics'.

India can be positioned as the hub for differentiated medicine as the country offers one of the most affordable development bases for personalized medicines. Personalized therapies will demand extensive clinical data generated from well-differentiated patient populations. India has one of the most desired disease and patient profiles that can enable such studies. Coupled with this is the need for a large number of novel diagnostics based on gene and non-gene based platforms. These are clearly large opportunities for Indian Biotech companies to pursue. Personalized drugs also address the affordability factor for expensive therapies such as those that are involved with cancer.

Some important barriers to improving the clinical utility of such knowledge exist. These include the highly complex nature of the problem, little incentive for industry to move to genomic-based approach, and lack of provider education.

Strategic Actions:

- Establish a cell for Diagnostic Biotechnology to encourage and support studies into the clinical application of pharmacogenomics. This cell should be well positioned to overcome barriers in its work to bring pharmacogenomics to the clinical setting.
- Encourage research-involving investigators with both clinical practice and pharmacology/pharmacokinetics expertise while at the same time keeping the overall goal of clinical application/utility in focus.
- Provide incentives for this group of clinician-researchers to bring these scientific advances to the patient bedside
- Support education programs to providers of the importance of this field and its utility.
- Encourage biopharmaceutical companies to include pharmacogenomic data in their drug submissions.

REGENERATIVE AND GENOMIC MEDICINE

The first wave of real healthy life extension therapies seems likely to result from research stem cells and regenerative medicine which helps natural healing processes to work faster, or uses special materials to regrow missing or damaged tissue. Doctors use regenerative medicine to speed up healing, and to help heal injuries that cannot heal on their own. Regenerative therapies have been demonstrated (in trials or the laboratory) to heal broken bones, bad burns, blindness, deafness, heart damage, nerve damage, Parkinson's and other conditions. Regenerative medicine will result in extended healthy lifespan; we will be able to repair some of the damage caused by aging, organ by organ. The first crop of simple stem cell therapies for regenerative medicine might be only a few years away from widespread availability. There are major scientific and ethical challenges and safety concerns that must be overcome in taking stem cell based technology for bench to bedside. As it is rapidly evolving field, the existing national (ICMR) guidelines need to be updated and supported by clear articulated procedures. India must consider the potential medical applications of stem cell research. We must reassure end users on the safety and quality by ensuring regulation on stem lines having stable characterizations so that safety risks are predictable. We must reassure suppliers by regulation from lab to market.

Strategic Actions:

- Formulate a comprehensive Human Tissue Act (end 2005) with codes and guidance for regenerative medicine. In the intension, ICMR and DBT will support existing guidelines for stem cell research with clear procedures to be followed by scientists and physician.
- DNA and stem cell banking facilities will be created.
- Lay down clearer laws on animal testing in the country for progress to be made in this sector.
- Emphasize on Intellectual Property Rights, confidentiality and feedback
- Regulation for human tissue engineered products.
- Public awareness to be created in order to allay fears through education programmes, industry conferences and seminars.

FOOD SECURITY

Food security refers to the availability of food and one's access to it. A household is considered food-secure when its occupants do not live in hunger or fear of starvation. According to the World Resources Institute, global per capita food production has been increasing substantially for the past several decades. In 2006, MSNBC reported that globally, the number of people who are overweight has surpassed the number who are undernourished – the world had more than one billion people who were overweight, and an estimated 800 million who were undernourished. According to a 2004 article from the BBC, China, the world's most populous country, is suffering from an obesity epidemic. In India, the second-most populous country in the world, 30 million people have been added to the ranks of the hungry since the mid-1990s and 46% of children are underweight.

Worldwide around 925 million people are chronically hungry due to extreme poverty, while up to 2 billion people lack food security intermittently due to varying degrees of poverty (source: FAO, 2010). Six million children die of hunger every year – 17,000 every day. As of late 2007, export restrictions and panic buying, US Dollar Depreciation, increased farming for use in biofuels, world oil prices at more than $100 a barrel, global population growth, climate change, loss of agricultural land to residential and industrial development, and growing consumer demand in China and India are claimed to have pushed up the price of grain. However, the role of some of these factors is under debate. Some argue the role of biofuel has been overplayed as grain prices have come down to the levels of 2006. Nonetheless, food riots have recently taken place in many countries across the world.

The ongoing global credit crisis has affected farm credits, despite a boom in commodity prices. Food security is a complex topic, standing at the intersection of many disciplines. A new peer-reviewed journal of Food Security: The Science, Sociology and Economics of Food Production and Access to Food began publishing in 2009. In developing countries, often 70% or more of the population lives in rural areas. In that context, agricultural development among smallholder farmers and landless people provides a livelihood for people allowing them the opportunity to stay in their communities. In many areas of the world, land ownership is not available, thus, people who want or need to farm to make a living have little incentive to improve the land.

In the US, there are approximately 2,000,000 farmers, less than 1% of the population. A direct relationship exists between food consumption levels and poverty. Families with the financial resources to escape extreme poverty rarely suffer from chronic hunger, while poor families not only suffer the most from chronic hunger, but are also the segment of the population most at risk during food shortages and famines. Two commonly used definitions of food security come from the UN's Food and Agriculture Organization (FAO) and the United States Department of Agriculture (USDA): Food security exists when all people, at all times, have physical, social and economic access to sufficient, safe and nutritious food to meet their dietary needs and food preferences for an active and healthy life.

Food security for a household means access by all members at all times to enough food for an active, healthy life. Food security includes at a minimum (1) the ready availability of nutritionally adequate and safe foods, and (2) an assured ability to acquire acceptable foods in socially acceptable ways (that is, without resorting to emergency food supplies, scavenging, stealing, or other coping strategies). (USDA)

The stages of food insecurity range from food secure situations to full-scale famine. "Famine and hunger are both rooted in food insecurity. Food insecurity can be categorized as either chronic or transitory. Chronic food insecurity translates into a high degree of vulnerability to famine and hunger;

ensuring food security presupposes elimination of that vulnerability. [Chronic] hunger is not famine. It is similar to undernourishment and is related to poverty, existing mainly in poor countries."

AGRICULTURAL AND FOOD SUPPLY CHAIN

The Indian agribusiness supply chain is highly fragmented, with independent players engaged in various value-addition activities such as input supplies, cultivation, processing, distribution, storage and retailing. The scale of operations of each entity is limited not only due to the structure of the supply chain, but also on account of Government policies as well as due to demand-related factors. In contrast, internationally, a large number of food companies are integrated across the chain. e.g. players such as ADM and Cargill have business interests in agricultural inputs, procurement, transportation and storage of produce, processing and value addition.

The structural complexion of the Indian supply chain thus translates into limited scale of financing as well as higher risk, given the lack of control of each of the players, on the supply chain.

ISSUES WITH FARMER FINANCING

It is important to understand the issues faced in farmer level financing, both from the borrower's as well as the lender's perspective, and develop solutions which address these issues comprehensively.

Limited Access to Credit

Non-availability of timely credit from organized sources, mainly on account of the processes and procedures of banks/financial institutions. This compels farmers to rely on intermediaries.

- Inadequacy in credit availability-The 'scale of finance' method, stipulated by RBI, caps total credit availability per farmer.
- Inability to offer tangible security/collateral cover to access credit from banks Subsistence farmers, who constitute a large proportion of India's farming community, are unable to benefit from the organized credit system since they cannot offer adequate tangible security. Further, the security offered (land documents) are notional and cannot be enforced by banks.
- The above factors, thus translate into continued reliance on intermediaries for financing. While intermediary financing is timelyas they do not require the level of documentation as banks, the cost of

funds is exorbitant, nearly 4 times that of interest rates charged by banks.

- A more efficient agri-supply chain requires that farmers' financing needs are addressed by banks and linkages of farmers with middlemen are scrapped.

Banker's Perspective

- Lack of credit discipline among borrowers Lack of credit discipline, to some extent is infused by the Government's decision to waive outstanding loans of farmers. Policy-related uncertainties translate into a huge risk for banks. Instead of loan waivers, the Government should seek to impose a short-term moratorium on repayments, to enable farmers to overcome the adverse impact of a poor monsoon/crop failure.
- Lack of tangible security from farmers-as mentioned above, farmers are unable to provide tangible security to financiers.
- Lack of linkages of farmers with processors-In absence of forward linkages with markets, banks run a huge performance risk, as there is greater potential for farmer default.
- Inadequate insurance cover the existing crop insurance schemes are inadequate, and this further increases the risk profile of farmers
- Norms for priority sector lending-As per the current RBI policy on priority sector lending, kharif credit is not accounted for in priority sector lending. This results in a lop-sided focus on extending financing to farmers, favouring crops cultivated in the rabi season.

Policy-related Issues

- Cooperatives Act: There are about 95,000 Primary Agriculture Credit Co-operatives (PACS) and about 100,000 marketing co-operatives. As per the Cooperative Act, cooperatives are allowed to borrow from Cooperative Banks, DCCBs and RRBs. This provision needs to be extended to scheduled banks, in order to enhance flow of credit to farmers.
- State Warehousing Corporations Act: Private sector scheduled banks are constrained from lending to State Warehousing Corporations. Given the sizeable investment required to create grain storage infrastructure, it would be in the interest of the CWC/SWCs that a wider participation of lenders is permitted.
- Restrictions on land holdings: The State Land Holding Laws prohibit consolidation of land holdings.

Financing to farmers with larger land holdings, and thus larger scale of operations, could result in higher credit flows from banks. Further, the

Government could also consider consolidation of wastelands and leasing of these to corporates. This would not only enable increase in financing to agriculture, but also aid crop diversification.

SOURCES OF FINANCE FOR AGRICULTURE & FOOD PROCESSING

Banks: The Food Processing sector has access to credit from Commercial Banks-Indian and Foreign, Cooperative Banks and the Regional Rural Banks. The formal institutional sector assists the food processing companies through long term loans for capital investments and short term loans for working capital. The total lending to the food processing sector is clubbed with lending to the agriculture sector and separate details for financing to the food processing sector are not available.

- National Bank for Agricultural and Rural Development (NABARD): NABARD offers refinance facilities for food processing, agri infrastructure, developmental assistance for RRBs, DCCBs, SSI and SHG linkages, and assists in infrastructure development and research.
- Small Industries Development Bank of India (SIDBI):SIDBI has been involved in assisting the entire SSI Sector including tiny, village and cottage industries through suitable schemes tailored to address the funding requirements for expansion, diversification, modernisation and rehabilitation.
- Export Import Bank (EXIM Bank): Exim Bank assists in financing and facilitation of foreign trade. The Bank also offers financial support to companies engaged in exports.
- National Cooperative Development Corporation (NCDC): NCDC assists in promotion, planning and financing the agricultural supply chain from production, processing, storage and trade of agricultural produce and food products. NCDC also provides assistance for marketing of certain notified commodities e.g. fertilizers, pesticides, agricultural machinery etc.
- Ministries/Government bodies
- Ministry of Food Processing Industries (MFPI): MFPI is the Nodal agency for development of the processed food sector in the country MFPI's financial schemes include schemes for technology upgradation, Human resource development, Quality testing, R&D,TQM, backward and forward integration, development of infrastructure including food parks.

- Agricultural and Processed Foods Products Export Development Authority (APEDA): APEDA facilitates market linkages between Indian producers, manufacturers and the international market. APEDA provides financial assistance for market development, infrastructure development and development of quality enhancing facilities.
- Ministry of Agriculture (MoA), Government of India: The Ministry of Agriculture under various schemes, provides financial assistance for development of specific crops for investment in seeds, irrigation, farm implements, inputs, infrastructure and training.
- National Horticultural Board (NHB):NHB promotes integrated development in horticulture, assists in development of post harvest management infrastructure, promotes production and processing of fruits and vegetables, strengthening of market information systems and assists in R&D programs in cultivation and processing. NHB's financial schemes are directed towards commercial horticulture and infrastructure related to post harvest techniques.

Financial assistance from these organizations are in the form of grants, back-ended subsidies, soft loans, refinance etc., with most of the schemes directed to specific sub-sectors of the agri/food processing industry. In addition, some State Governments offer financial schemes to companies in the sector. The key issues with Government schemes are highlighted below:

- Overlap among schemes: Various Government agencies have separate schemes which aim to achieve the same objectives. Thus, an applicant can potentially approach different agencies and obtain assistance from more than one agency. Some of the examples of overlapping schemes are as follows:
- Several areas addressed inadequately: There are a host of areas which are not addressed by these schemes such as financing of freezer cabinets for retail outlets, financing of bulkcoolers for milk, financing of alternate markets/mandis, financing of vending machines for tea/coffee/beverages etc.
- Creation of excess capacity: There are instances where assistance has been provided to set up cold storages in locations which have excess storage capacity. This has lead to an unviable situation for all cold storage operators in the region. It is essential for the nodal agencies to achieve development of the food processing sector in a sustainable manner.
- Inadequate financial assistance^ most instances, there is a cap on total financial assistance provided at INR 5-5.5 mn. This leads to fragmentation in capacity creation, and often, a situation of overcapacity. At the same time, it does not allow players to scale up their operations.

- Inadequate Monitoring of progress of schemes: It is important to track progress of the projects funded by the Government. Periodic appraisal of the progress will ensure that funds are appropriately utilized. This will also enable assessment of success and lacunae of these schemes and record the multiplier effect of the financial support provided.
- Most of the schemes are back-ended which leads to a funds crunch during implementation.
- Institutions providing subsidies do not undertake their own assessment and rely on the appraisal report of the bank which provides loans to the project.
- There are significant time lags from the date of application for financial assistance, to release of funds, and may affect the project schedule, and lead to cost overruns
- These schemes do not address working capital requirements, which due to the nature of the industry are significant and often higher in terms of quantum than capital investment requirements
- Applicants claim that they are often unaware that their applications have been rejected on account of lack of communication from the funding institution. Further, applicants are unaware of the reasons for rejection.

Sources of Finance for Agriculture

At present nearly 60% of rural credit is used for agriculture and allied activities, 10% for non-farm activities and the balance 30% for funding household consumption expenditure. Nearly 40-45% of rural credit needs are catered to by formal credit institutions and the balance by the informal sector including commission agents, input suppliers, traders, Self Help Groups (Joint Liability Groups), processing industries and professional money lenders. The Indian commercial banks are mandated by the Reserve Bank of India (RBI), to undertake directed lending for agriculture and rural development. Commercial banks are required to achieve priority sector lending of 40% of net bank credit of which 18% should be for agriculture. Further, sub-targets are also specified i.e., Direct Agriculture (credit for direct farmer benefit, Target: Minimum of 13.5% of the net bank credit) and Indirect Agriculture (Target: Maximum of 4.5% of the net bank credit).

In FY 03, total agricultural advances outstanding under priority sector lending, by Public Sector Banks stood at INR 735 billion, on a total net bank credit portfolio of INR 4779 billion. As compared to this, agricultural advances by Private Sector Banks stood at INR 119 billion, on a total net bank credit portfolio of INR 718 billion. The following Exhibit provides details of direct and indirect priority sector credit amongst Public and Private sector banks during FY03. In case of a shortfall in lending to agriculture, banks are required to invest in Rural Infrastructure Development Fund (RIDF) deposits, which offers interest rates linked to the bank's performance in lending to agriculture. These rates are inversely proportional to the shortfall in agricultural lending. The interest rate on RIDF is provided in slabs, based on the achievement of the priority sector lending target, and can be as low as 4% to 5%.Therefore, it is imperative to achieve the agriculture target and more specifically the Direct Agriculture target as it accounts for 75% of the total.

Established in 1996, RIDF is being utilized for making investments in rural infrastructure projects. Nine tranches of RIDF have been established with an aggregate corpus of INR 340 billion, with the ninth tranche of INR 55 billion. Cumulative sanctions and disbursements under various tranches of RIDF stood at INR 334 billion and INR 193 billion respectively, at the end of February 2004.The RIDF funds are used to finance minor irrigation projects, shallow wells, drinking water facilities, roads, bridges and storage infrastructure.

AMENDMENT OF RESTRICTIVE ACTS/POLICIES

As mentioned earlier, the following Acts need to be amended to facilitate increased flow of funds to the agriculture and food processing sectors

- Cooperatives Act-Allow scheduled banks to finance cooperative societies.
- State Warehousing Corporations Act-Allow scheduled banks to finance CWC/SWCs
- Land Ceiling Act allow consolidation of land holdings to enable large scale cultivation
- Further, banks should be dispensed with the need for documentation for hypothecation of crops for loans upto INR 0.5 mn. These advances should be considered secured on the basis of a letter of undertaking
- Stamp Duty: The stamp duty charged on agricultural loans payable by farmers is an unnecessary burden on farmers, while enhancing Government revenues only marginally and should be abolished.

Banks/financial Institutions

- Corporate/market linkages: Banks/financial institutions need to factor in linkages of processors with companies/markets for assured off take, as also backward linkages for procurement

- Corporate security: Banks/financial institutions can seek security from output buyers (secondary processor/marketer) such as a "put option "for a specified period to take off the loan or a part of the loan in case of default
- Track record in risk assessment Banks need to allocate higher weightage to the track record of food companies while providing financing to them. This includes market feedback on company management and its operations, relationships with suppliers and buyers etc. Such market feedback will also provide insights on management's ability to deal with cyclically.
- Leverage parameters: Most food processing companies have high leverage as the net worth of the companies is low and the current liabilities are high. A higher leverage ratio can be considered for food processing companies, keeping in view the nature of the business, as compared to companies in other industry segments
- Cash Collateral as Security: Cash collateral in the form of cash or cash equivalents which can be converted into cash without any erosion in value, can be taken as security. With the company providing a cash collateral, a higher weightage is assigned for this factor, as the risk associated with the loan is mitigated to a large extent. A cash collateral factor can supplement the profitability factor e.g. low weightage can be assigned to Profit After Tax (PAT) in the case of availability of high collaterals.
- Size criteria: Most companies in the food processing sector have limited scale of operations, with turnover in the range of INR 100 million to INR 150 million, on account of the structure of the supply chain as described earlier. The company's operation cannot be assessed based on turnover. Thus, low weightage should be assigned to the turnover criteria.

Role of State Governments

State Governments should consider participation with the industry in the food processing sector through minor equity participation in private projects, which lends a 'State project status'. This can not only be beneficial in terms of facilitating speedy clearances for projects, but also lend state machinery (unused land/state-owned storage facilities/state extension services) to these projects.

4

Principles of Food Production

Food Hygiene Includes all Practices, Precautions and Procedures Involved in:

a. Protecting food from the risk of biological, chemical or physical contamination.

b. Preventing any organisms multiplying to an extent that would expose consumers to risk or result in premature decomposition of food.

c. Destroying any harmful bacteria in food by thorough cooking or processing.

Benefits from high standards of food hygiene include:

a. Reduced risk of food poisoning, foreign body contamination and spoilage.

b. Compliance with MOD and legal requirements.

c. Economic advantages, including increased shelf life and reduction of waste.

d. Consumer satisfaction and enhanced reputation.

e. Increased morale of personnel.

Hazard Analysis

The Food Safety (General Food Hygiene) Regulations 1995 (FS(GFH)R) refer to *Hazard Analysis* (HA). HA is based on five principles, detailed at Annex A to this chapter and are summarised below:

a. Analysis of the potential food hazards in a food operation.

b. Identification of the points in those operations where food hazards may occur.

c. Deciding which of the points identified is critical to food safety ("critical points").

d. Identification and implementation of effective control and monitoring procedures at those critical points.

e. Review of the analysis of food hazards, the critical control points and the control and monitoring procedures periodically and whenever the catering operation changes.

In order to meet the requirements of HA across the MOD, a practical guide for the MOD has been produced, based on the principle that there are four main methods of preparing food. The guide, *Hazard Analysis-The Four Line Method,* is at Annex B to this chapter, which is to be used as a template in all Service establishments. Universally recognised Critical Control Points (CCPs) have been identified, however additional Control Points or Critical Points may be applicable in each case. This will be entirely dependant on the type of catering function and the supporting infrastructure. When preparing individual HA profiles, units are advised to consider the following definitions in order to determine the type of risk that applies at each stage of food preparation:

a. *Control point* – specific stages in those operations where food hazards may occur.

b. *Critical point* – control points which are considered critical points where the hazard must be controlled to ensure that it is eliminated or reduced to a safe level.

c. *Critical control point* – is a critical point in the process where a specific hazard must be controlled on the basis that no further process will adequately eradicate that hazard.

Written records are also important in establishing a Due Diligence defence under the Food Safety Act 1990. HA, as defined in the FS(GFH) R, includes the principles of: Critical Points and the Control of those Critical Points. In this document the term Hazard Analysis (HA) will be used, but this is be taken to be synonymous with HACCP.

Regulation 4(3) of FS(GFH)R states that: "a proprietor of a food business shall identify any step in the activities of the food business which is critical to ensuring food safety and ensure that adequate safety procedures are identified, implemented, maintained and reviewed". This Regulation is designed to make caterers focus on the activities critical to food safety and to find ways of controlling them.

The systematic analysis of each individual food item would result in a heavy burden for most MOD catering operations and would be impractical to implement. An approach, that considers the catering operation step by step, from supply through to consumption, is achievable, effective and forms the basis of food HA in the MOD. The HA approach to ensuring food safety and hygiene is intended to give a clearer focus on the controls that are important to Service caterers to ensure that safe food is provided.

Where catering is provided by contract, the contractor is to use a Hazard Analysis system that meets legislative requirements or adopt the Four-Line Method. Specialist advice, through the Chain of Command may be obtained from DCG Service Support Contracts

Supply of Food and Water

The FS(GFH)R state that no raw materials or ingredients shall be accepted by a food business if they are known to be contaminated with parasites, pathogenic micro organisms, toxic, decomposed or foreign substances and that, after normal sorting and/or preparatory processing procedures, they would still be unfit for human consumption.

Delivery Monitoring Routine checks must be made on deliveries of food for signs of damage, contamination and the presence of pests. The general condition of the food is to be checked, together with more specific checks such as date marks (Best Before and Use By) and temperature. Unfit food or food past its "Use By" date must not be accepted and is to be immediately returned. Temperature monitoring should be conducted at the time of delivery and should form part of the quantity & quality check. For chilled and frozen foods, checks are to be made that the food is delivered at the correct temperature (less than 8oC for chilled and less than –12oC for frozen). The Delivery Monitoring form is no longer required to be completed, the vehicle temperature printed record must now be attached to the Price Advice Note (PAN). For bulk stock "between pack" temperatures are to be taken using an air probe and recorded directly onto the PAN. A note stating "Quality & Date Coding Checked" is to be hand written onto the PAN. A commercially produced ink stamp could also be utilised for this task.

Potable Water Supply. It is essential that galleys and kitchens be supplied with potable water for the purposes of food preparation and cleaning.

Food Storage

The FS(GFH)R requires food products and ingredients to be stored in food premises in appropriate conditions, designed to prevent harmful deterioration and to protect food from contamination.

Dry Provisions Storage. These storerooms are to be kept clean and orderly to minimise the potential hazards from "foreign bodies" and to prevent the harbourage of pests. Where practicable these stores are to be proofed against pest ingress. Part used packs are to be adequately to prevent contamination.

High ambient temperatures (above 13oC) and high humidity are to be avoided. Dry goods should be stored on suitable racking raised off the ground.

Chilled Storage (5oC). Cool rooms and refrigerators are to be kept clean and tidy. It is the responsibility of the Catering Manager to raise the appropriate request if refrigerated space is considered inadequate. Best Practice guidance indicates that the ideal operating temperature of a refrigerator is 1-4oC.

Frozen Storage(-18—21oC). Freezers are to be kept clean and tidy at all times. Best Practice guidance indicates that the ideal operating temperature of a freezer is between –18oC to –21oC.

Satisfactory storage is essential if a galley/kitchen/mess is to serve clean and safe food. This should include routine turnover of stock and checks of Best Before Dates. There are four main groups of foods that require differing storage conditions. The groups are as follows:

Where reference is made to a Priced Advice Note; this also means any other form of Invoice or Delivery Note

a. *Fresh Fruit and Vegetables.* These may be contaminated by soil bacteria and are to be stored away from other foods in a cool area, with adequate ventilation, preferably refrigerated at a temperature appropriate to the product.
b. *Dry Foods.* (including Canned Foods, Cereals and Flour). These are to be stored in dry, well-ventilated rooms ideally at 13oC. Food is to be put on racks, shelves or pallets off the floor.
c. *Frozen Foods.* These are to be stored in a freezer as soon as they are delivered. Frozen foods are to be used within 1 month. 'Kitchen prepared' chilled foods that are subsequently frozen are also to be used within 1 month.
d. *Perishable Foods.* Perishable food temperatures in a refrigerator must not exceed the legal limit of 8oC. Therefore, perishable foods are to be stored in a refrigerator cabinet, set at a target temperature of 5oC. Such food includes chilled dishes, meat, poultry, eggs, cooked meats and vegetable dishes, prepared salads, soft cheeses, sandwiches, fresh pasta, low acid desserts, smoked or cured meats, fish and dairy produce. Best Practice advice, is that chilled foods are to be used within 48hrs. Food that is spoiled or past its "Use By Date" must be removed from food rooms and appropriate action taken prior to disposal. Broken packaging must be avoided.

Food Displays/Serveries. Food displays and serveries should be set at the correct temperature with adequate time allowed for the equipment to achieve target temperatures. It is good practice to have sneeze screens attached to hot and cold food displays to minimise the risk of contamination of food. The handles of utensils should not contact food; this can be achieved by using utensils with longer handles than the service containers.

Cross Contamination

Cross contamination is the process whereby pathogenic bacteria present in raw food, such as meat, poultry, and vegetables, come into contact with ready to eat food, either directly or indirectly, through contact with surfaces, equipment, splashes, drips, utensils, hands and cloths. To prevent cross contamination, raw and ready to eat foods are to be kept apart. Good management planning of workflow through the kitchen can assist in this. The flow of waste is also be considered, along with raw products and prepared food. Additionally, it is to be ensured that:

a. Where possible, there are separate designated preparation areas for raw and cooked foods. If, due to limitation of space, preparation surfaces are used for both high and low risk foods then these activities must be separated by time (ideally low risk foods prepared first). Preparation surfaces must be thoroughly cleaned and disinfected between the two operations.

b. If raw and cooked food is to be stored in the same refrigerator, raw foods are always to be stored (and adequately covered) below cooked or salad foodstuffs.

c. Food handlers always wash their hands between handling raw and cooked foods and change their protective clothing once heavily soiled.

d. Separate colour coded cutting boards (red for raw meat, blue for raw fish, green for fresh fruit, salad and vegetables, white for cooked foods) and colour-coded knives are used (details of NSNs for these items are contained in JSP 308). Always thoroughly wash and disinfect cutting boards and knives in between and after use.

e. Separate machines for raw and cooked food are used, or slicing and mincing machines are thoroughly cleaned and disinfected between raw and cooked food. If separate machines are not available, cooked food must be prepared before raw food.

f. All food is kept in covered, dated and labelled containers during storage and before service.

g. Storing and Cooking with Eggs. Some eggs can contain salmonella bacteria inside or on their shells so care must be taken with their use. They are to be stored in a cool dry place or refrigerated, with exterior packaging removed and kept away from other foods. All egg dishes are to be served immediately or cooled quickly and chilled. Commercially pasteurised egg products should be used in preparation of any recipe that is served uncooked or partially cooked. This includes mayonnaise; bearnaise and hollandaise sauces, some salad dressings, ice-cream, icings and tiramisu.

h. Washing of Fruits and Vegetables. All fruits and vegetables should be washed in potable water before use. In addition fruits and vegetables purchased locally overseas are to be immersed for 30 minutes in water containing either of the agents listed. The items should then be rinsed to remove taste of disinfecting agent:
   (i) Calcium hypocholrite granules-one level teaspoon per two gallons of water
   (ii) Milton: 5 teaspoons per gallon of water.

DEFROSTING FOOD

It is essential that frozen meat and poultry are thoroughly thawed on trivets, or if unavailable in a designated defrosting area, or ideally in a defrosting cabinet. Frozen vegetables must be cooked directly from frozen.

Cooking

It is essential that food is cooked thoroughly to destroy any bacteria on or within it. Food is to be cooked to a core temperature of 75°C, checked with a calibrated digital probe (disinfected with bactericidal wipes). It is important that food such as rolled joints, burgers, and chicken are cooked to 75°C at its thickest part.

Chilling Food

When chilling cooked food for later cold or hot consumption, it is essential that the food be cooled as quickly as possible. In accordance with 'Best Practice' hot food is to be taken through the recognised 'Danger Zone' of 60-10ºC within 4hrs, and subsequently stored in a refrigerator at 8ºC or cooler.

(These requirements refer to the temperature of the food, not the air temperature of the equipment).

If the target temperatures are not achieved food is to be discarded. The chilling of food to a recognised safe temperature may be achieved using the following methods:

a. Break food down into shallow trays and cover.

b. Cool initially to a low ambient temperature, then refrigerate or blast chill to below 'Danger Zone'.

c. Store in a refrigerator.

Reheating Food

It is advised that the reheating cooked foods is avoided, but if this is not possible it must only be reheated once and the following steps are to be taken:

a. Ensure that the food reaches a core temperature of at least 75°C. (In Scotland it is a legal requirement for the centre of the food being reheated to reach 82°C).

b. Use digital probe thermometers to check the temperature at the centre of the food. (Probes are to be disinfected using bactericidal wipes between each use).

c. The food is to be served and eaten as soon as possible.

d. After reheating, any leftover cooked food is to be thrown away.

Temperature Control and Monitoring

The Food Safety (Temperature Control) Regulations 1995 (FS(TC)R) impose two holding temperatures of below 8ºC or above 63ºC. The purpose of the regulations is to inhibit or prevent harmful micro organisms from multiplying by keeping food outside of the recognised 'Danger Zone' of 60ºC to 10ºC. The FS(TC)R allow limited periods outside temperature control during preparation, display, service, storage or transport, but it is an offence to keep food out of temperature control for so long that it could become unsafe. The requirements do not apply to foods, which by their nature or their packaging (e.g. canned foods) are inherently safe from the growth of such organisms. In such cases the manufacturer's guidelines are to be followed.

Refrigerator and Freezer Temperature Recording. The operating temperature of all appliances should be monitored and recorded thrice daily using either of the templates at Appendices 1 and 4 to Annex B of this Chapter. As a guide, the target operating temperature for a refrigerator is 5°C and for a freezer is-18°to-21°C, but this will depend on the efficiency, age, contents and capacity of the appliance.

Note: The template to be utilised is dependent on establishment size and the number of appliances.

Catering Standing Orders are to specify the corrective action (including Out of Hours/Weekend actions required) to be taken in the event of refrigerators/freezers not achieving recommended operating temperatures or failing completely.

CURRENT LEGISLATION

Registration and Licensing of Premises

To exercise control over premises used for food handling the responsible agency must know the location of the premises, and its task of ensuring that these premises are properly constructed and adequately equipped, and that satisfactory hygienic standards are maintained, is facilitated if a business is not permitted to open until the premises are officially approved. This power of control is strengthened if the agency is able to enforce the closure of unhygienic premises.

Many countries recognize the need to control the opening of catering premises so as to ensure that before business commences they comply with the standards in force. Provision is usually made also for the closure of premises in the event of serious contraventions of the law.

The opening of premises is controlled in various ways. Sometimes a permit or licence from the local health department is required, and this may be revoked or suspended if the premises are allowed to fall into an unsatisfactory condition. One large city requires an applicant for a permit to complete successfully a course of instruction in food protection conducted by the health department. Sometimes there is a system requiring premises to be approved before they are opened.

In some countries control is achieved by a requirement that all catering establishments must be registered. For practical purposes no real distinction can be drawn between licensing and registration although licensing is normally for a limited period and implies periodical renewal of licences. However, the health authority will need for its own purposes a recordof the premises for which it is responsible, and a registration scheme therefore seems most appropriate.

WHO Technical Report Series, No. 439, 1970

Registration of food establishments is the process by which food control authorities, local or national, maintain registers in which is entered information that applicants must supply before they may operate. This information should include the name and address of the establishment and other relevant data. The approval of premises may be required before registration; however, this practice is not universal. Registration is usually for an indefinite period of time, subject to cancellation for causes indicated in the law. Licensing of food establishments is the process by which appropriate food control authorities, local or national, issue licences to operators of food establishments that have complied with relevant legal requirements. The licence authorizes the establishment to operate for a specified period of time, after which renewal is required. Licences may be suspended or revoked as provided for in the law.

Where licensing or registration schemes are not operated the task of the health authorities is more difficult, but they could still have power to enforce the closure of premises that present a danger to health. The requirement that premises may not be opened for business until they have been registered or licensed is of value to both the health authorities and the operator of the business.

If catering premises may be opened without any prior consultation with the health authority, or without their approval, the operator may be faced subsequently with the need to make substantial alterations at heavy expense in order to comply with the food hygiene regulations. Such expense might be avoided if the operator-is required to consult the health authorities at the outset and accepts guidance from health officers in the design and equipping of the premises. Indeed, the health authorities could warn operators against attempting to use an unsuitable building that could not be. made to comply with the required standards.

Authority to enforce the closure of insanitary premises is most .important. Clearly, there must be a means of speedily closing catering establishments that constitute a danger to public health, but the machinery for doing this requires much consideration.

Construction

Regulations for the construction and equipping of catering premises in force in different countries vary considerably in detail although their basic intention is to provide conditions under which food can be prepared hygienically and without endangering the health of the consumer. There can be little dispute about the basic constructional elements required, or about the essential pieces of equipment necessary for good food hygiene. It is obvious, however, that there is a wide difference of opinion on the extent to which it is necessary to impose on operators of catering businesses detailed legislative requirements concerning the buildings in which they carry on their business and the equipment to be installed.

Regulations may, in addition to specifying some matters in detail, state certain broad principles. For example, one set of national regulations for food premises contains this provision:

"Every food premises shall be so constructed, located and maintained that,

(a) the premises are free from every condition that may,

  (i) be dangerous to health,

  (ii) injuriously affect the wholesomeness of food prepared, processed, packaged, served or stored therein; ..."

Some other regulations have a similar general requirement worded as follows: "No food business shall be carried on at any insanitary premises or place or at any premises or place the condition, situation or construction of which is such that food is exposed to the risk of contamination."

A Further Example Reads

"Any person who sells, prepares, packages or stores for sale any food under insanitary conditions shall be guilty of an offence." General provisions of this kind are desirable because in legislation it is difficult to specify in detail all conditions which should constitute an offence. The extent to which legislation contains detailed requirements regarding the construction of catering premises varies considerably. The points usually covered are that in rooms where food is stored or prepared the floors must be of smooth, nonabsorbent materials and so constructed as to be easily cleaned; all walls and ceilings in such rooms must be capable of being readily cleaned; all parts of these rooms must be kept in good order, repair and condition so as to enable them to be effectively cleaned. Premises are often required to be of such construction and in such a state of repair as to resist the entry of rats, mice and insects. In some causes, openings to the outer air are required to be effectively protected against the entry of flies and other flying insects.

Lighting and Ventilation

It is usual to require all food preparation and storage rooms to be adequately lighted and ventilated. This is normally stated in general terms but in some areas the degree of illumination is specified in detail. For example, one set of rules prescribes that at least 215 lux (20 foot candles) of light must be provided on all working surfaces and equipment in food preparation, utensil washing, and hand washing areas and in toilet rooms. At least 54 lux (5 foot candles) at a distance of 75 centimetres (30 inches) are required in dining rooms and all other areas during cleaning operations.

Equipment and Utensils

It is usual for food hygiene legislation to contain provisions concerning equipment and utensils. Legislation may require equipment and utensils to be of such materials, workman-ship and design as to be smooth, easily cleaned and resistant to damage. Provisions are sometimes laid down that all surfaces that come into contact with food must be readily accessible for cleansing and inspection. It is quite common to impose an obligation to ensure that all articles, equipment and surfaces must be nontoxic and, as far as possible, nonabsorbent. In one country legislation requires equipment and utensils to be:

(a) of sound and tight construction;

(b) kept in good repair;

(c) of such form and material that it can be cleaned and disinfected;

(d) free from cracks, crevices and open seams; and

(e) corrosion-resistant and nontoxic.

A most important part of the operation of catering establishments is the cleansing of equipment and utensils. The duty of ensuring that these are effectively cleansed after use and maintained in a clean and sanitary condition may be imposed in very general terms, leaving it to the operator of the business to adopt whatever methods he regards as the most effective. In such cases the legislation may simply require facilities for the cleansing of utensils and equipment to be provided and utensils and equipment to be thoroughly cleansed after use and kept clean. On the other hand, regulations in force in some countries specify in considerable detail how the cleansing of utensils and equipment should be carried out.

Sometimes, sinks with two or three compartments must be provided if cleansing is performed manually. Such detailed regulations usually divide cleansing into three operations: prerinsing, cleansing, and disinfecting or sanitizing. The temperature of the water used for cleansing may be laid down, 43.5°C (110°F) being usual. For machine washing, a higher water temperature may be specified.

Use of a Detergent may also be Obligatory

There may be a requirement that after utensils have been washed they should be sanitized by immersion either in hot water or in water containing a sterilizing agent. The temperature of the water used for sanitizing is usually required to be between 76.5°C and 82°C (170°-180°F) with an immersion time of 30 seconds at the lower temperature or 10 seconds at the higher. Where methods of chemical sterilization are specified the required immersion time varies from 45 to 60 seconds. The sterilizing agents allowed are solutions containing 50-100 parts per million of available chlorine, 12.5-25 parts per million of available iodine, or quaternary ammonium compounds at a concentration of not less than 200 parts per million. These solutions are to be used at a temperature of not less than 24°C (75°F).

Rules are also sometimes laid down for the operation of dishwashing machines. Where these are of the immersion type the. requirements are similar to those for manual dishwashing. For spray-type machines the temperature of the rinsing and sanitizing water may be specified. Detailed instructions may be given for cleansing procedures. These may require that, before being washed, articles must be flushed or scraped, and when necessary soaked, to remove large food particles and soil. Effective concentrations of a suitable detergent may be insisted upon for both manual and mechanical dishwashing.

As a check on the efficiency of. dishwashing, utensils may have to meet a prescribed bacterial standard. This is usually a plate count of not more than 100 bacterial colonies per utensil, using the swab technique.

In the design of legislation for the cleansing of utensils and equipment there are clearly two different philosophies. The first is that it is sufficient merely to require the provision of adequate facilities for washing, such as hot and cold water and sinks, and to make it obligatory to cleanse all vessels and other equipment thoroughly after use.

The operator of the food business is left to decide for himself the precise methods to be used to achieve a satisfactory standard of cleanliness. The second philosophy is that not only must there be an obligation to provide the necessary cleansing facilities, but the methods to be used must be prescribed in considerable detail including the temperature of the hot water and procedures for sterilization.

This philosophy also maintains that visual cleanliness is not enough; after being washed and sterilized, equipment must satisfy a bacteriological test. The need for strict cleanliness is unquestionable and the law should be adequate to enforce this. It is doubtful, however, whether there is really any need to embody in legislation the precise methods and materials to be used. If failure to cleanse equipment adequately is made an offence with an appropriate penalty for noncompliance, this should give the health officer the necessary authority to ensure that proper cleansing procedures are followed. His training should enable him to advise on how the required standards can be maintained.

Washing Facilities for Staff

All food premises need adequate washing facilities for the staff. They must be provided with a sufficient number of suitable wash-hand basins for the use of all persons engaged in the handling of food on or around the premises, and the basins should be conveniently placed in an accessible position. Wash-hand basins should be provided with an adequate supply of hot and cold water or of hot water at a controlled temperature. Soap or other suitable detergent, nailbrushes, and clean towels or other drying facilities must also be supplied. These washing facilities should not be used for any purpose other than personal hygiene. Food hygiene regulations generally contain provisions of this kind. Some insist that signs must be displayed conspicuously, instructing workers to wash their hands after using toilets.

Sanitary Facilities

Satisfactory sanitary facilities are necessary in all food premises for the use of the staff and this is usually legally enforceable. Sometimes, sanitary accommodation is also specifically required in catering establishments for customers' use. This accommodation must be adequately lighted and ventilated and kept clean and in good order.

Sleeping Places

It is usual for food hygiene legislation to prohibit the unhygienic practice of using rooms in which food is prepared, served or stored for sleeping purposes, and in some cases the legislation bans direct communication between sleeping quarters and rooms in which food is handled.

Refuse

Some regulations make it obligatory to provide adequate space, in a suitable situation, for the removal of waste from food, the separation of food fit and unfit for human consumption, and the storage of waste and unfit food before disposal. There must be no avoidable deposit or accumulation of solid or liquid wastes in a food room. Sometimes the law is more precise. It may insist, for example, that all garbage and refuse in kitchen areas should be kept in separate, leakproof, nonabsorbent containers equipped with tight-fitting covers unless otherwise protected from flies and other insects. There may be a specific obligation to maintain refuse storage areas in a clean sanitary condition. Garbage and refuse should be disposed of as often as necessary to prevent decom-position or overflow.

First-aid

The provision of first-aid equipment for food handlers is obligatory in some countries and should be generally compulsory. Storage of clothing It is usual to require lockers to be provided for the outdoor or other clothing and footwear of employees arriving for work in food handling establishments.

Water Supply

An adequate supply of clean and wholesome water is generally stipulated for all food premises. Clearly, this should be a universal requirement.

Food Protection

It is common to embody in food hygiene legislation an obligation to protect food from contamination by insects, rodents, dust, and from all other kinds of contamination. In addition, some countries specifically require the exclusion of domestic animals and birds from premises in which food is handled commercially. In some cases provision is made to avoid contamination of food by toxic materials such as insecticides, rodenticides, and other substances used for the maintenance of hygienic conditions.

The points covered are labelling of containers of toxic materials so that the contents may be easily identified, storage of these substances in locked cupboards, and their use in such a manner as not to contaminate food or endanger health.

Food hygiene regulations also often contain obligations concerning the control of the temperature of food in catering establishments. These controls usually apply to perishable or potentially hazardous foods, that is, to foodstuffs that support the growth of pathogenic organisms.

The general principle embodied in this legislation is that the foodstuffs concerned should not be kept any longer than is necessary at a temperature favourable to the growth of bacteria. There is some variation in the temperature range specified but the extreme limits are 4.4°C (40°F) and 65.5°C (150°F). This range, with slight variations, is regarded as a danger zone to be avoided as far as possible.

The foods to which temperature control applies are described in different ways. In some cases the legislation refers to "perishable food" or "perishable food of animal or vegetable origin". Sometimes particular foods are referred to, such as "custards, cream fillings or similar products". In one case regulations apply to "infection and toxin prone food" which is defined as "perishable food consisting in whole or in part of milk, milk products, meat, poultry, fish, shellfish or any other ingredient capable of supporting the rapid growth of pathogenic organisms or the production of the toxins of such organisms".

In another case temperature control applies to all food "consisting of meat, fish, gravy, or imitation cream, or prepared from, or containing, any of these substances or any egg or milk". There are, however, exceptions for such articles as bread, biscuits and cakes containing egg or milk as ingredients that were introduced before baking, chocolate or sugar-confectionery, butter and other fats, cheese and uncooked bacon and ham. It is not usual to impose legal requirements for the cooking of food but some regulations provide that pork products must be thoroughly cooked to heat all parts to a minimum temperature of 65.5°C (150°F). Stuffings,-poultry, stuffed meats and stuffed poultry must be heated throughout to a minimum temperature of 74°C (165°F).

Food Handlers

It is a standard requirement of food hygiene legislation that food handlers must be clean and practise hygienic habits. They are generally required to wear clean, washable overclothing and, in some countries, to wear headgear that confines the hair. Smoking in food preparation areas is often prohibited. In some cases food handlers are specifically required to wash their hands after visits to the toilet. One set of regulations requires employees and management to be adequately informed concerning acceptable and sanitary food handling practices. They are expected to attend approved food handler training programmes when these, are available. Another set of regulations requires supervisors of food processing establish-ments, including caterers, restaurants and eating places, to complete a course in food protection. Many countries in their food hygiene legislation try to ensure that persons who become infected and who may transmit infection to food are excluded from handling food while they are in this condition or require all food handlers to undergo periodical medical examinations to determine that they are healthy and-free from infection.

For example, some regulations for restaurants require any restaurant licensee who suspects that an employee has contracted any form of communicable disease, or has become a carrier of such disease, to exclude the employee from the restaurant and notify the local health officer immediately. The health officer has to determine whether the food handler is infected or is a carrier, and laboratory examinations may be required. No persons who have a communicable disease or are carriers of such disease are permitted to work in a restaurant. Restaurants are forbidden to employ such persons or any persons suspected of being so affected. Chest X-ray examinations and/or tuberculin tests are recommended annually for all restaurant personnel. Persons who at any time have had typhoid or paratyphoid fever may not be employed in a restaurant until it has been definitely determined that they are not carriers.

Similar regulations elsewhere provide that every food handler must be free from, and not a carrier of, a disease that could spread through the medium of food, and that food handlers must submit to such medical examinations and tests as the medical officer of health or the minister of health may require. No person who has a communicable skin disease or infection or who resides in a dwelling where there is communicable disease is permitted work as a food handler, unless-he obtains a certificate from the medical officer of health showing that he is free from infection.-The operator of a food business who knows or suspects that an employee is violating these provisions must notify the medical officer of health.

Another example of this kind of legislation is an order prohibiting persons liable to taint or contaminate food from engaging in the handling of food.-This provision applies, inter alia, to persons who:

(1) are suffering, or suspected to be suffering, from typhoid or paratyphoid fever or other Salmonella infections, dysentery, or staphylococcal infections;

(2) are carriers of the microorganisms causing these diseases;

(3) show detectable clinical symptoms of infectious hepatitis;

(4) are suffering, or suspected to, be suffering, from tuberculosis in a communicable state;

(5) are suffering, or suspected to be suffering, from a contagious disease of the skin.

This legislation also requires all persons professionally engaged in the production or distribution of foodstuffs or substances used as food and who are in direct contact with such products to undergo an annual examination to verify that they are free from tuberculosis in a communicable state.

Legislation in another country places an obligation on any person employed as a food handler and who knows that he, a member of his family, or another person who lives in the same dwelling is suffering from a dangerous disease to inform his immediate superior of the facts. The latter must immediately notify the health board. '

This legislation also provides that no person who is suffering from boils, a rash or sores, or from a dangerous disease, or from tuberculosis, leprosy or syphilis, may be permitted to engage in the handling of foodstuffs intended for sale. This also applies to carriers of disease. Where necessary, the person concerned must produce a medical certificate proving that the disease from which he is suffering is not contagious or liable to cause food poisoning.

Further, before any person is engaged, or begins to work, as a food handler he must produce a medical certificate to show that he is not suffering from a disease which, having regard to the nature of the work, could involve a danger of contagion or food poisoning through the agency of foodstuffs. The medical certificate must not have been issued more than 30 days previously. It must be kept and, on request, shown to the health and police authorities.

In yet another country, immediately a person engaged in the handling of food becomes aware he is suffering from, or is a carrier of, typhoid or paratyphoid fever or any other Salmonella infection or amoebic or bacillary dysentery or any staphylococcal infection likely to cause food poisoning, he is required to inform his employer who must immediately notify the proper officer of the local authority. Infected persons may be ordered to discontinue or to refrain from engaging in any occupation connected with the preparation and handling of food or drink for human consumption until further notice is given by the local authority that the risk of transmitting infection no longer exists. Food handlers suspected of being carriers of any of the diseases mentioned above may also be required to undergo a medical examination.

One local authority may require food workers, before employment, or at any other time, to undergo a chest X-ray examination or any other test or examination deemed necessary for the protection of the public health. Certification of such examinations may be issued by the local health authority and, when required by the authority, a copy of the certificate for each employee must be in the possession of the management of the food service establishment.

Certificates are not Valid for more than 12 Months from the Date of Issue

Certain countries require food handlers to hold a health booklet or certificate. In one of these countries workers who are to be employed in the food industry are obliged to undergo a pre-employment medical examination. This examination is necessary even for temporary employees, apprentices, and students. The workers covered by this obligation include those employed in the sale of foodstuffs and the provision of meals to the public, as well as in the distribution of snacks and refreshments.

Also in that country, food workers must undergo other preventive medical examinations at times determined by the hygiene services. In addition, they must undergo, without delay, emergency medical examinations:

(a) if they are suffering from diarrhoea, a purulent disease or a fever, or a communicable disease, or if they are suspected or suffering from a communicable disease;

(b) if a communicable disease, or a suspected case of such a disease, occurs in the work-place or at home.

These medical examinations include a detailed clinical examination, a medical history, and a microbiological examination, particularly with a view to the detection of tuberculosis and possible carriers of the pathogens of typhoid and paratyphoid and other salmonelloses and shigelloses.

Workers engaged in the food industry are obliged to hold a health booklet issued by the health community medical officer for the area in which they live or, if they work in a factory having a medical officer, by that officer. In the case of workers employed for short periods a certificate as to their medical fitness is issued in place of the booklet. The booklet (or certificate) is kept by the person in charge of the undertaking, but where workers carry out their work outside the undertaking, they are required to carry the booklet (or certificate) themselves. The person in charge of the undertaking must keep a list of the workers indicating when they were medically examined, the results of the examination, and when the next medical examination should be carried out.

At each medical examination the worker must produce his booklet or certificate. If the physician finds, or suspects, that the worker is suffering from a disease or infection he must take charge of the booklet or certificate and give the worker a written receipt in exchange. The worker is obliged to hand over the receipt, without delay, to the person in charge of the undertaking. Every food establishment must ensure that the handling of food is carried out only by workers holding a booklet or certificate; work may not be commenced before the booklet or certificate has been issued. The establishment must notify the health community medical officer in good time of every worker who is required to undergo a preventive medical examination.

Another country imposes compulsory medical examinations on certain groups of workers including those in catering establishments. In addition to the usual examinations for typhoid, paratyphoid, dysentery and other Salmonella infections, workers in catering establishments are subject to examination for infestation with intestinal parasites. A worker who is transferred to another undertaking, establishment or installation is not subject to a prior medical examination if his health booklet shows that his initial pre-employment examination was satisfactory, that his state of health has been under systematic surveillance, and that during the previous two years, no member of his family or other person living in his household has suffered from an infectious intestinal disease or an intestinal parasitosis. After every absence from work lasting more than one month the person concerned must be examined as if he were beginning work for the first time. Every food handler (i.e., any person who carries out, directly or indirectly, manual activities involving foodstuffs in such a manner as to be able to affect their wholesomeness, quality or hygiene) in another country must undergo the health examinations prescribed by the health authority for the purpose of acquiring a health certificate entitling him to work in that occupation. The certificate must be renewed annually or at more frequent intervals should the health authority consider it necessary. Any person who, while not holding the appropriate health certificate, engages in food handling and the proprietor of the establishment where he works are both liable to penalties.

Another example which may be mentioned is an ordinance concerning hygiene in restaurants and catering establishments. Every worker must undergo a pre-employment medical examination to determine his suitability for the work he will be engaged in. The nature of the examinations required, the classes of workers subject to such examinations, and the frequency of, and procedure for, the medical examinations are prescribed.

A recent piece of legislation on food hygiene states that only persons who provide a medical certificate indicating that they are free from diseases or infections that could be transmitted to other persons by means of foodstuffs may commence work involving the handling of foodstuffs other than pre-packed foodstuffs, or work directly related to hygiene in the handling of foodstuffs, in various types of premises including restaurants and other catering establishments. This legislation also requires persons working in the types of premises to which it applies to undergo at least once a year a medical examination or other health checkup as officially prescribed. Physicians or veterinarians who have good grounds for believing that food poisoning has been caused by a foodstuff contaminated by pathogenic bacteria or extraneous substances during food handling operations are required to notify the local medical officer.

BASIC PRINCIPLES OF FOOD HYGIENE LEGISLATION

General

An essential part of the health services of every country must be to ensure that food is safe, wholesome and not injurious to health, and that proper hygienic standards are observed at all stages from production or manufacture until it reaches the consumer. The requirement that food must not be injurious to the consumer means that food traders must be under a legal obligation not to sell food which they know to be unfit for human consumption or could with reasonable diligence ascertain to be unfit. They must also have a duty to take due care to ensure that adequate standards of food hygiene are observed at all times in the conduct of their business.

It is not sufficient merely to place legal obligations on food traders. There must also be official agencies charged with the duty of exercising close surveillance of all types of food premises, including catering establishments. Many environmental factors can influence the maintenance of satisfactory standards of food hygiene and, as indicated on p. 3, progress will be made more easily if a food hygiene programme is developed within a progressive environmental health and personal health programme. Where food hygiene is the responsibility of the agency responsible for general environmental health there should be no problems in securing coordination of effort. Where separate agencies are responsible it is important that there should be close liaison between them. Agencies with the duty of controlling food hygiene should not be regarded merely as law enforcement bodies, though they should possess adequate legal powers to establish and maintain the required standards. Their officers need to have authority to enter food premises during business hours without warning in order to carry out the necessary inspections and investigations, and it should be an offence to obstruct officers who are carrying out these duties.

Effective legislation is necessary to secure the proper observance of hygienic practices. The basic legal requirements should be embodied in national statutes but the more detailed and technical requirements should be contained in regulations which can be readily and quickly changed to keep pace with scientific and technical developments. There is a need also for codes of practice which, while not legally enforceable, give guidance to food traders and the officials responsible for inspection and control of food premises on equipment, materials and practices for maintaining the desired standards.

Although the basic legal requirements embodied in national statutes and the more detailed and technical requirements contained in regulations will cover very many points it is not advisable that legal provisions should be too rigid or too detailed. Often, the more satisfactory approach is to prescribe the objective to be achieved and to allow some flexibility in the choice of methods to be used to achieve that objective. In any case, it is impracticable for laws to be drafted in sufficient detail to take account of all possible circumstances.

Basic Legal Requirements

The principal points that should be included in the basic legal requirements are as follows :

(a) the authorities responsible for the enforcement of food hygiene legislation should be specified;

(b) it should be an offence to sell food that is diseased, unsound, or unwholesome;

(c) recognized officers of the responsible authorities should be authorized to inspect food intended for sale, to take samples, and to seize and condemn food that appears to be unfit for human consumption;

(d) recognized officers should have power of entry to all food premises;

(e) all food premises of prescribed types should be required to be registered by the responsible authorities before being opened for business and these authorities should be empowered to revoke registration where continued use of the premises constitutes a potential danger to health; particular types of premises would be registered under regulations made in accordance with point (f) below;

(f) authority should be given to the appropriate minister to make regulations containing specific and detailed legal requirements with respect to food hygiene in particular types of premises and different food handling activities; compliance with the appropriate regulations would be a condition for the registration of premises;

(g) physicians who become aware, or suspect, that patients they are attending are suffering from food poisoning should be required to notify the local public health authority;

(h) authority should be given to recognized officers to prohibit the use or removal from the premises of food which they believe is likely to cause food poisoning until investigations have shown whether it may safely be used for human consumption;

(i) persons engaged in the handling of food should be required to notify their employers if they become aware that they are suffering from any condition that might lead to the infection of foods and employers should be under an obligation to inform the local medical officer.

Model Food Hygiene Regulations for Catering Establishments

The basic legal requirements for the control of food hygiene which should be embodied in a national statute have been set out in the preceding section.

As indicated in point (f), detailed legal requirements should be stated in regulations or secondary legislation made under the authority given by the statute. It should be noted that in these model regulations expressions such as "adequate", "suitable", and "sufficient" are frequently used.

This is because a legal document of this kind must have sufficient flexibility to meet the varying circumstances found in different types of establishment and in different parts of the world. The interpretation of these terms is left to the discretion of local control authorities.

The following Model Code details those matters which it is suggested should be included in the food hygiene regulations for catering establishments.

Registration

(a) No premises shall be used as catering premises unless they are registered for that purpose by the local authority. Any person who uses unregistered premises shall be guilty of an offence.

(b) Application for registration must be made before the premises are opened and the application must be refused if the premises do not comply in all respects with food hygiene regulations for catering premises.

(c) Registration may be revoked if at any time serious breaches of the food hygiene regulations occur.

Where a licensing system is preferred, points (e) and (f) should be amended accordingly.

Where a system of licensing is preferred this section should be amended accordingly.

Construction and Maintenance of Premises

(a) No catering business shall be carried on in any premises in which food is exposed to the risk of contamination.

(b) The premises shall be of adequate size, sound construction and in good repair. Areas used for the handling of food shall be completely separated from any part of the premises used as living quarters.

(c) The internal walls, floors, windows, ceilings, woodwork and all other parts of every room shall be so constructed and maintained in such good order, repair and condition as to enable them to be cleaned effectively and to prevent, as far as practicable, infestation by rats, mice, insects and birds.

(d) The premises must be kept free from rodent and insect pests, and birds, and dogs, cats and other domestic animals must be excluded from areas where food is processed or stored. All rodenticides, insecticides and other toxic substances must be clearly labelled and, when not in use, be kept in separate, locked cupboards.

(e) The premises must be kept in a clean condition and no water must be deposited or allowed to accumulate in rooms in which food is prepared, stored or served, except in so far as this may be unavoidable for the proper conduct of the business.

(f) All rooms in which food is stored, prepared or served, or in which utensils are washed, and all sanitary conveniences shall be well ventilated and well lighted.

(g) No room in which food is stored, prepared or served may be used as a sleeping place or communicate directly with a sleeping place or sanitary convenience.

Water Supply

(a) An ample supply of water of drinking quality must be provided at all premises; where possible this should be from the public supply main but if from any other source it must be approved by the health authority and checked for bacteriological and chemical quality at intervals.

(b) An adequate supply of hot and cold water or of hot water at a suitable controlled temperature shall be provided in all areas where food is prepared and where equipment and utensils are washed.

(c) Water used for making ice shall be of drinking quality.

Drainage

(a) All premises shall be provided with an adequate drainage system which, where practicable, shall be connected to the public sewerage system.

(b) Where connexion to the public sewerage system is not practicable the method of sewage disposal must be approved by the health authority.

Personal Washing Facilities

(a) An adequate number of suitable wash-hand basins shall be provided for the use of all persons engaged in the handling of food. These basins must be conveniently accessible to these persons and in any case must be located in or immediately adjacent to all sanitary conveniences. In all new premises, and where practicable in existing premises, hand-washing facilities must be provided within or immediately adjacent to food preparation areas. All wash-hand basins shall be properly connected to the drainage system and be provided with an adequate supply of hot and cold water or of hot water at a suitable controlled temperature. At each basin shall be provided an adequate supply of soap or other suitable detergent, nail brushes and clean towels or other suitable drying facilities.

(b) Facilities for personal washing must not be used for any other purpose.

Sanitary Facilities

(a) An adequate number of conveniently located sanitary conveniences shall be provided for all employees. Separate conveniences shall be provided for each sex. An adequate supply of toilet paper must be available at all times.

(b) In or near every sanitary convenience a clearly legible notice must be displayed in a prominent position requiring users to wash their hands after using the convenience.

Storage of Clothing

Adequate facilities shall be provided for the storage of employees' clothing and personal belongings not being worn during working hours and such clothing and belongings shall not be kept on the premises other than in the accommodation provided.

First-aid Materials

(a) A sufficient supply of suitable bandages, dressings (including waterproof dressings) and antiseptics shall be provided for the first-aid treatment of persons engaged in the handling of food.

(b) This material must be kept in suitable locations readily accessible to employees and clearly labelled.

Storage of Refuse

(a) A sufficient number of watertight containers of durable and nonabsorbent material with tight-fitting or self-closing lids shall be provided for the storage of food waste and other refuse.

(b) Adequate space shall be provided for the temporary storage of waste; this space must not be in any room where food is stored, prepared or served.

Food

(a) Food, while being stored, prepared, displayed or served, shall be protected from dust, flies, rodents and other pests, and other contamination and all persons engaged in the handling of food shall take all reasonable steps to protect the food from risk of contamination.

(b) Raw meat, poultry, fish, shellfish, milk or uncooked products containing those foodstuffs shall not be permitted to come into contact with any cooked foods. Any persons handling any of these raw or uncooked foodstuffs must, before handling any cooked food, thoroughly wash their hands and carefully cleanse all surfaces, utensils and equipment with which the foodstuffs may have come into contact.

(c) All perishable food consisting in whole or in part of milk or milk products, eggs, meat, poultry, fish, shellfish or other ingredients capable of supporting the rapid growth of pathogenic organisms shall be kept at a temperature that is not within the range from 4.5°C (40°F) to 65.5CC (150°F), except during necessary periods of preparation and service.

(d) Raw, unprocessed fruits and vegetables shall be thoroughly washed in clean drinking-water before use.

(e) All ice used for cooling drinks or food by direct contact shall be made from water of drinking quality.

Equipment

(a) All articles of equipment with which food comes into contact, or may come into contact, must be kept clean, be so constructed, placed, of such materials, and kept in such good order, repair and condition as to:

(i) enable them to be thoroughly cleaned;

(ii) prevent, so far as is reasonably practicable, any matter being absorbed by them; and

(iii) prevent, so far as is reasonably practicable, any risk of contamination of the food.

(b) An adequate number of sinks or other suitable facilities for the washing of food and equipment when necessary shall be provided. These sinks or other facilities shall be provided with an adequate supply of hot and cold water or of hot water at a suitable controlled temperature, or with cold water only where the sink is used only for washing fish, fruit or vegetables. These facilities must not be used for any other purpose.

(c) Equipment shall be so installed as to facilitate its cleaning and the cleaning of adjacent areas.

Personnel

(a) Every person handling or coming into contact with food, or with any equipment or utensil used in its preparation, processing or service shall:

   (i) keep as clean as may be reasonably practicable all parts of his body that are liable to come into contact with the food and, in particular, he shall wash his hands thoroughly before commencing work and after using a sanitary convenience;

   (ii) wear clean, washable outer garments and headgear that confines his hair;

   (iii) keep any open cut or abrasion on any exposed part of his body covered with a suitable waterproof dressing; and

(iv) refrain from spitting and using tobacco in any form while engaged in the handling, preparation or service of food, while handling utensils or equipment, and while he is in any room containing open food.

(b) No person who is affected with any disease in a communicable form may work in any catering premises and the proprietor of any catering establishment shall not permit any such person to work in the establishment. Immediately a person engaged in the handling of food becomes aware, or has reason to suspect, that he is suffering from, or is a carrier of, typhoid, paratyphoid or any other Salmonella infection or amoebic or bacillary dysentery or any staphylococcal infection likely to cause food poisoning he shall immediately inform the person operating the catering premises in which he is employed and that person shall immediately notify the medical officer of the health authority.

FOOD & BEVERAGE DEPARTMENT-RESTAURANT DIVISION

Out of the three divisions that the Food & Beverage Department holds, the Restaurant Division is the most popular when it comes to money. Its structure has changed recently, but it's fairly simple and everyone knows who to report to. Described in a simple way, the structure goes like this:

Restaurant Operations Manager

He or she is a three stripe officer and is in charge with the entire restaurant operation. The Windjammer Manager, Portofino Manager and Assistant Dining Room Manager report directly to him or her and, of course, the position is got its own admin now.

The contract length is 4 months with 2 months off and, aside the thousands of dollars salary, he or she is got monthly bonuses from china and linen savings, wine gratuity and the big, fat bonus at the end of the fiscal year.

Windjammer Manager

The Windjammer Manager is a two and a half stripes officer and is in charge with Windjammer Buffet, Johnny Rockets (fast food outlet) and Café Promenade (coffee shop with snacks). Directly to him reports the following positions: Johnny Rockets Supervisor and Windjammer Supervisor.

Just like the Restaurant Operations Manager, the contract length is also 4 months on with 2 months off, and a big, fat bonus included.

Portofino Manager

Portofino Manager is in charge with the Italian specialty restaurant. Is a two stripes officer with a 6 months contract and sharing cabin. From 2 stripes below, no officer is entitled to the yearly bonus, but they got part of the cover charge. The Portofino waiters report directly to him.

Assistant Dining Room Manager

This position was introduced recently. It's a 2 stripe position with a contract length of 6 months and sharing cabin.

Their salary is based on tips, but don't be fooled by it, they make a lot of money. There are 3 Assistant Dining Room Managers, each in charge with one Dining Room deck (our Dining Room has 3 decks/levels). They are responsible with the booking on their respective decks and the Headwaiters working in that deck report directly to them.

Restaurant Operation Admin

The admin reports directly to the Restaurant Operations Manager and basically does everything. He has 2 office boys/girls helping in the morning with paperwork. Is a one stripe officer with 6 months contract and sharing cabin.

Head Waiter

These guys are making more money than a 3 stripe officer because their salary is based on tips. They are one stripe officers, with 6 months contract length and a sharing cabin. Each of them has a station assigned (there are 3 stations on each Dining Room deck) and the waiters and assistant waiters working there report to him/her.

Johnny Rockets Supervisor

He or she is in charge with Café Promenade and Johnny Rockets and reports to the Windjammer Manager. This is a one stripe officer with a 6 months contract and sharing cabin. They get a fix salary with no gratuities. The Café Attendants and Johnny Rockets Attendants report to him/her.

Windjammer Supervisor

There are 2 Windjammer Supervisors onboard, both reporting to the Windjammer Manager. They are in charge with supervising the Windjammer operation and the Assistant Waiters working there are reporting to them. It's a one stripe position, with fix salary, 6 months contract and sharing cabin.

Room Service Supervisor

The Room Service Supervisor reports directly to the Restaurant Operations Manager and is in charge with Room Service operation and Attendants working there. Is a one stripe position, with 6 months contract, fix salary and sharing cabin.

Room Service Operator

The position is pretty much self-explanatory. They are part of the crew with a 6 months contract, fix salary and sharing cabin. The Room Service Operator reports to Room Service Supervisor.

Mess Supervisor

The Mess Supervisor is in charge with both officer and crew messes and directly to him/her are reporting the Mess Attendants. Mess Supervisor is a one stripe position with a 6 months contract, fix salary and sharing cabin. He or she reports to Restaurant Operations Manager.

Room Service Attendant

The above 5 positions are similar; the only difference is the place they work in. Except for the Johnny Rockets who get gratuities from the cover charges in the fast food outlet they work in, the rest have a fix salary and they also get gratuities from the beverages they sell (not too much though). They all have an eight months contract and sharing cabin.

Assistant Waiter

This is a much better position that the ones above because they have a salary based on tips. An average assistant waiter salary can be around $2000 or more a month. Also, the contract length is 6 months and they only work in the Dining Room or Windjammer during the dinner time and in Windjammer (during the day time).

During the diner time, they work 4 weeks in the Dining Room and one week in Windjammer. They get gratuities in the Dining Room, but in Windjammer they get only the gratuities from the beverages they sell. Might not make sense, but this is how the system is set up.

Waiter

Waiter position is a good position as well due to the amount of money they make and the 6 months contract length. An average waiter makes around $3000 and, during the diner time, works only in Dining Room. As a first-timer, this structure might look confusing. Indeed there are a lot of positions in the Restaurant Division, but this is what makes it easier because everyone knows exactly what they are suppose to do.

FOOD AND BEVERAGE SERVICE OCCUPATIONS

Significant Points

- Most jobs are part time and many opportunities exist for young people—nearly 2 out of 3 food counter and fountain workers are 16-19 years old.
- Job openings are expected to be abundant through 2008, reflecting substantial turnover.
- Tips comprise a major portion of earnings; consequently, keen competition is expected for bartender, waiter and waitress, and other jobs in popular restaurants and fine dining establishments where potential earnings from tips are greatest.

Nature of the Work

Whether they work in small, informal diners or large, elegant restaurants, all food and beverage service workers aim to help customers have a positive dining experience in their establishments. These workers are responsible for greeting customers, taking food and drink orders, serving food, cleaning up after patrons, and preparing tables and dining areas. All of these duties require a high quality of services customers will return.

The largest group of these workers, *waiters* and *waitresses,* take customers' orders, serve food and beverages, prepare itemized checks, and sometimes accept payments. Their specific duties vary considerably, depending on the establishment where they work. In coffee shops, they are expected to provide fast and efficient, yet courteous service.

In fine restaurants, where gourmet meals are accompanied by attentive formal service, waiters and waitresses serve meals at a more leisurely pace and offer more personal service to patrons. For example, servers may recommend a certain wine as a complement to a particular entree, explain how various items on the menu are prepared, or complete preparations on a salad or other special dishes at table side. Additionally, waiters and waitresses may check the identification of patrons to ensure they meet the minimum age requirement for the purchase of alcohol and tobacco products.

Depending on the type of restaurant, waiters and waitresses may perform additional duties usually associated with other food and beverage service occupations. These tasks may include escorting guests to tables, serving customers seated at counters, setting up and clearing tables, or operating a cash register. However, formal restaurants frequently hire other staff to perform these duties, allowing their waiters and waitresses to concentrate on customer service.

Bartenders fill drink orders that waiters and waitresses take from customers. They prepare standard mixed drinks and, occasionally, are asked to mix drinks to suit a customer's taste. Most bartenders know dozens of drink recipes and are able to mix drinks accurately, quickly, and without waste, even during the busiest periods.

Besides mixing and serving drinks, bartenders collect payment, operate the cash register, clean up after customers leave, and often serve food to customers seated at the bar. Bartenders also check identification of customers seated at the bar, to ensure they meet the minimum age requirement for the purchase of alcohol and tobacco products. Bartenders usually are responsible for ordering and maintaining an inventory of liquor, mixes, and other bar supplies. They often form attractive displays out of bottles and glassware and wash the glassware and utensils after each use.

The majority of bartenders who work in eating and drinking establishments directly serve and interact with patrons. Because customers typically frequent drinking establishments for the friendly atmosphere, most bartenders must be friendly and helpful with customers.

Bartenders at service bars, on the other hand, have little contact with customers because they work in small bars in restaurants, hotels, and clubs where only waiters and waitresses serve drinks. Some establishments, especially larger ones, use automatic equipment to mix drinks of varying complexity at the push of a button. Even in these establishments, however, bartenders still must be efficient and knowledgeable in case the device malfunctions or a customer requests a drink not handled by the equipment.

Hosts and *hostesses* try to create a good impression of a restaurant by warmly welcoming guests. Because hosts and hostesses are restaurants' personal representatives, they try to insure that service is prompt and courteous and that the meal meets expectations. They may courteously direct patrons to where coats and other personal items may be left and indicate where patrons can wait until their table is ready. Hosts and hostesses assign guests to tables suitable for the size of their group, escort patrons to their seats, and provide menus. They also schedule dining reservations, arrange parties, and organize any special services that are required. In some restaurants, they also act as cashiers.

Dining room attendants and *bartender helpers* assist waiters, waitresses, and bartenders by cleaning tables, removing dirty dishes, and keeping serving areas stocked with supplies. They replenish the supply of clean linens, dishes, silverware, and glasses in the dining room and keep the bar stocked with glasses, liquor, ice, and drink garnishes. Bartender helpers also keep bar equipment clean and wash glasses. Dining room attendants set tables with clean tablecloths, napkins, silverware, glasses, and dishes and serve ice water, rolls, and butter. At the conclusion of meals, they remove dirty dishes and soiled linens from tables. Cafeteria attendants stock serving tables with food, trays, dishes, and silverware and may carry trays to dining tables for patrons.

Counter attendants take orders and serve food at counters. In cafeterias, they serve food displayed on counters and steam tables, carve meat, dish out vegetables, ladle sauces and soups, and fill beverage glasses. In lunchrooms and coffee shops, counter attendants take orders from customers seated at the counter, transmit orders to the kitchen, and pick up and serve food. They also fill cups with coffee, soda, and other beverages and prepare fountain specialties, such as milkshakes and ice cream sundaes. Counter attendants prepare some short-order items, such as sandwiches and salads, and wrap or place orders in containers for carry out. They also clean counters, write itemized checks, and sometimes accept payment.

Fast-food workers take orders from customers at counters or drive-through windows at fast-food restaurants. They pick up the ordered beverage and food items, serve them to a customer, and accept payment. Many fast-food workers also cook and package food, make coffee, and fill beverage cups using drink-dispensing machines.

Working Conditions

Food and beverage service workers are on their feet most of the time and often carry heavy trays of food, dishes, and glassware. During busy dining periods, they are under pressure to serve customers quickly and efficiently. The work is relatively safe, but care must be taken to avoid slips, falls, and burns.

Part-time work is more common among food and beverage service workers than among workers in almost any other occupation. Those on part-time schedules include half of all waiters and waitresses, and 6 out of 10 food counter and fountain workers, compared to 1 out of 6 workers throughout the economy. Slightly more than half of all bartenders work full-time with 35 percent working part-time and the remainder working a variable schedule.

The wide range in dining hours creates work opportunities attractive to homemakers, students, and other individuals seeking supplemental income. In fact, nearly 2 out of 3 food counter and fountain workers are between 16 and 19 years old. Many food and beverage service workers work evenings, weekends, and holidays. Some work split shifts—that is, they work for several hours during the middle of the day, take a few hours off in the afternoon, and then return to their jobs for evening hours.

Employment

Food and beverage service workers held over 5.4 million jobs in 1998. Waiters and waitresses held about 2,019,000 of these jobs; counter attendants and fast-food workers, 2,025,000; dining room and cafeteria attendants and bartender helpers, 405,000; bartenders, 404,000; hosts and hostesses, 297,000; and all other food preparation and service workers, 280,000.

Restaurants, coffee shops, bars, and other retail eating and drinking places employed the overwhelming majority of food and beverage service workers. Others worked in hotels and other lodging places, bowling alleys, casinos, country clubs, and other membership organizations. Jobs are located throughout the country but are typically plentiful in large cities and tourist areas. Vacation resorts offer seasonal employment, and some workers alternate between summer and winter resorts, instead of remaining in one area the entire year.

Training, Other Qualifications, and Advancement

There are no specific educational requirements for food and beverage service jobs. Although many employers prefer to hire high school graduates for waiter and waitress, bartender, and host and hostess positions, completion of high school is usually not required for fast-food workers, counter attendants, and dining room attendants and bartender helpers. For many people, a job as a food and beverage service worker serves as a source of immediate income, rather than a career. Many entrants to these jobs are in their late teens or early twenties and have a high school education or less. Usually, they have little or no work experience. Many are full-time students or homemakers. Food and beverage service jobs are a major source of part-time employment for high school and college students.

Because maintaining a restaurant's image is important to its success, employers emphasize personal qualities. Food and beverage service workers are in close contact with the public, so these workers should be well-spoken and have a neat, clean appearance. They should enjoy dealing with all kinds of people and possess a pleasant disposition. Waiters and waitresses need a good memory to avoid confusing customers' orders and to recall faces, names, and preferences of frequent patrons. These workers should also be good at arithmetic so they can total bills without the assistance of a calculator or cash register if necessary. In restaurants specializing in foreign foods, knowledge of a foreign language is helpful. Prior experience waiting on tables is preferred by restaurants and hotels that have rigid table service standards. Jobs at these establishments often have higher earnings, but they may also have higher educational requirements than less demanding establishments.

Usually, bartenders must be at least 21 years of age, but employers prefer to hire people who are 25 or older. Bartenders should be familiar with State and local laws concerning the sale of alcoholic beverages.

Most food and beverage service workers pick up their skills on the job by observing and working with more experienced workers. Some employers, particularly those in fast-food restaurants, use self-instruction programs with audiovisual presentations and instructional booklets to teach new employees food preparation and service skills. Some public and private vocational schools, restaurant associations, and large restaurant chains provide classroom training in a generalized food service curriculum.

Some bartenders acquire their skills by attending a bartending or vocational and technical school. These programs often include instruction on State and local laws and regulations, cocktail recipes, attire and conduct, and stocking a bar. Some of these schools help their graduates find jobs.

Due to the relatively small size of most food-serving establishments, opportunities for promotion are limited. After gaining some experience, some dining room and cafeteria attendants and bartender helpers are able to advance to waiter, waitress, or bartender jobs. For waiters, waitresses, and bartenders, advancement usually is limited to finding a job in a more expensive restaurant or bar where prospects for tip earnings are better. A few bartenders open their own businesses. Some hosts and hostesses and waiters and waitresses advance to supervisory jobs, such as maitre d'hotel, dining room supervisor, or restaurant manager. In larger restaurant chains, food and beverage service workers who excel at their work are often invited to enter the company's formal management training program.

Job Outlook

Job openings are expected to be abundant for food and beverage service workers. Employment of food and beverage service occupations is expected to grow about as fast as the average for all occupations through 2008, stemming from increases in population, personal incomes, and leisure time. While employment growth will produce many new jobs, the overwhelming majority of openings will arise from the need to replace the high proportion of workers who leave this occupation each year. There is substantial movement into and out of the occupation because education and training requirements are minimal, and the predominance of part-time jobs is attractive to people seeking a short-term source of income rather than a career. However, keen competition is expected for bartender, waiter and waitress, and other food and beverage service jobs in popular restaurants and fine dining establishments, where potential earnings from tips are greatest.

Projected employment growth will vary by type of food and beverage service job. Growth in the number of families and the more affluent, 55-and-older population will result in more restaurants that offer table service and more varied menus—requiring waiters and waitresses and hosts and hostesses. Employment of fast-food workers also is expected to increase in response to the continuing fast-paced lifestyle of many Americans and the addition of healthier foods at many of these restaurants. However, little change is expected in the employment of dining room attendants, as waiters and waitresses increasingly assume their duties. Employment of bartenders is expected to decline as drinking of alcoholic beverages outside the home—particularly cocktails—continues to drop.

Earnings

Food and beverage service workers derive their earnings from a combination of hourly wages and customer tips. Earnings vary greatly, depending on the type of job and establishment. For example, fast-food workers and hosts and hostesses usually do not receive tips, so their wage rates may be higher than those of waiters and waitresses and bartenders, who may earn more from tips than from wages. In some restaurants, these workers contribute a portion of their tips to a tip pool, which is distributed among the establishment's other food and beverage service workers and kitchen staff. Tip pools allow workers who normally do not receive tips, such as dining room attendants, to share in the rewards of a well-served meal. In 1998, median hourly earnings (not including tips) of full-time waiters and waitresses were $5.85. The middle 50 percent earned between $5.58 and $6.32; the top 10 percent earned at least $7.83. For most waiters and waitresses, higher earnings are primarily the result of receiving more in tips rather than higher hourly wages. Tips usually average between 10 and 20 percent of guests' checks, so waiters and waitresses working in busy, expensive restaurants earn the most.

Full-time bartenders had median hourly earnings (not including tips) of $6.25 in 1998. The middle 50 percent earned from $5.72 and $7.71; the top 10 percent earned at least $9.19 an hour. Like waiters and waitresses, bartenders employed in public bars may receive more than half of their earnings as tips. Service bartenders are often paid higher hourly wages to offset their lower tip earnings.

Median weekly hourly earnings (not including tips) of full-time dining room attendants and bartender helpers were $6.03 in 1998. The middle 50 percent earned between $5.67 and $7.11; the top 10 percent earned over $8.49 an hour. Most received over half of their earnings as wages; the rest of their income was a share of the proceeds from tip pools.

Full-time counter attendants and fast-food workers, except cooks, had median hourly earnings (not including tips) of $6.06 in 1998. The middle 50 percent earned between $5.67 and $7.14, while the highest 10 percent earned over $8.45 a hour. Although some counter attendants receive part of their earnings as tips, fast-food workers usually do not.

In establishments covered by Federal law, most workers beginning at the minimum wage earned $5.15 an hour in 1998. However, various minimum wage exceptions apply under specific circumstances to disabled workers, full-time students, youth under age 20 in their first 90 days of employment, tipped employees, and student-learners. Employers are also permitted to deduct from wages the cost, or fair value, of any meals or lodging provided. However, many employers provide free meals and furnish uniforms. Food and beverage service workers who work full time often receive typical benefits, while part-time workers usually do not.

In some large restaurants and hotels, food and beverage service workers belong to unions—principally the Hotel Employees and Restaurant Employees International Union and the Service Employees International Union.

Perspectives on Productivity

Most studies of productivity are drawn from manufacturing industries, however, services require an approach that recognises their peculiar characteristics. In the provision of services the processes of production and consumption are often simultaneous or can overlap-generating a specific set of challenges (Johnston & Jones, 2004; Mullins, 1998). Customers are active participants in many service encounters, and therefore an important influence on productivity (Martin et al., 2001).

Quality is so closely entwined with more measurable outcomes in service provision that it becomes very difficult to isolate any one influence on productivity. The intangibility of service, and therefore the importance of psychological outcomes (e.g. comfort and pleasure) in the process of quality creation, represent major challenges in measuring and understanding service sector productivity in general, and the F&B sector in particular.

This section of the report reviews the perspectives held by both employers and employees on productivity in their NZ F&B work-places and the factors that influence it. The thematic areas addressed below reflect the core themes raised by the 20 interviewees and the focus groups employee participants. Survey responses are also reviewed.

What is Productivity and how is it Measured?

The issue of how to measure productivity in the service sector is a critical one for this study. Gummesson (1998) suggests that measurements of service productivity can be 'ambiguous and inadequate', resulting in unhelpful comparisons between industries.

Food and beverage service providers often focus on reducing labour costs while maintaining sales (labour generally comprises around a third of hospitality costs, and food or beverage products, another third) (Reynolds, 2004). Concentrating on the reduction of labour costs as a percentage of sales may achieve short term productivity targets but can also jeopardise long term viability due to the erosion of service standards. Poor service affects customer satisfaction, which in turn influences sales and productivity, thereby creating a cycle of poor productivity.

Some commentators argue that a primary focus in F&B service productivity is quality. Parasuraman (2002) argues that how a company allocates service inputs will directly affect productivity from the customer's perspective. This view is upheld by Bates, Bates and Johnston (2003) who found that providing better service, while labour intensive, yields higher long-term profits per employee. They concluded that better service providers have significantly better return on equity and return on total assets than their counterparts with poorer service levels.

Measuring productivity in hospitality is acknowledged to be particularly difficult (Atkinson & Brown, 2001). David, Grabski and Kasavana (1996) found nine different measurements of productivity alone in their survey of hotel chain bench-marking. Most food service contract companies use sales per labour hour to express productivity (Clark, 1997a), but according to Reynolds, do not measure the cost of food sales, which he views as a 'good measure of labour productivity, but not of operational performance' (1998).

The heavy use of part-time and seasonal labour by F&B enterprises adds a further level of complexity to measuring productivity in the sector.

The situation in NZ mirrors the international context quite closely. It is clear that several managers and employees do not have a good understanding of productivity and what it means in the context of their business. As the interviews and focus groups progressed, however, it became obvious that factors related to, and influenced by, productivity were very much in people's minds-its just that productivity was not always the catchall term used to help think about them.

Both the interviews and survey show that managers view customer satisfaction and loyalty rate as the most important factors to use in assessing productivity (74% and 50% of survey respondents), whereas financial measures (meeting budget, costs and balancing, and revenue/sales per person) are 'most important' indicators for under a third of respondents.

Table : Indicators used to assess productivity

	Most important (%)	Least important (%)
Customer Satisfaction	74	11
Return visitation / loyalty	50	12
Volume	30	30
Meeting Budget	27	27
Cost and Balancing	24	45
Revenue / Sales per person	18	41
Number of table turns	13	50
Covers Per Staff	0	67

Larger F&B enterprises (especially those housed in hotels) have fairly sophisticated financial targets on which productivity is measured.

"[Productivity is] the amount of product that gets prepared and cooked and sold in relation to the number of staff I have... how many man hours for our front of house staff measured against revenue for a particular night".

Labour costs and food costs are the most critical and closely reviewed areas underpinning productivity. Fast food/takeaway outlets were found to be acutely aware of the various 'actual' versus 'desired' ratios they are working to.

Nevertheless for most managers (regardless of size or type of enterprise) productivity is defined by customer satisfaction. Favourable feedback from customers through comment cards, tips, and unsolicited remarks are viewed as vital indicators of productivity.

"Productivity is about how customers feel. The quality of the feedback we get from customers in terms of positive comments and tips".

"Communication between customers and staff-observation of their relationships tells me about the level of quality, and therefore, productivity".

Measurement of customer satisfaction ranged from sophisticated feedback forms (larger accommodation based operations) through to the simple observation of client behaviour, responses and tipping (smaller operations). In many cases smaller operators are going by their "gut-feelings" and the number of regulars that return with friends and new clientele are a critical 'measure' of productivity.

Productivity is also correlated with staff being able to meet or exceed expectations in a team setting.

"Defining and measuring productivity must be seen in terms of teamwork. Individual performance leads to team work. And if that works, it flows smoothly. And if it doesn't, we can then isolate with whom the problems lie".

Contented staff and management capability are the two factors management deem to have the most positive influence on productivity, followed by 'internal communication', 'training' and 'employee retention'. The survey results were very much reflecting the information gathered from the interviews-with managers across size and type of enterprise all stressing the importance of staff happiness, communication, and their own ability to manage these situations effectively.

Table : Factors impacting on productivity (% of respondents)

	Negative influence (%)	*Positive influence (%)*	*Mean**
Contented staff	0	85	4.9
Management capability	0	83	4.8
Internal communication	0	62	4.6
Training	2	70	4.5
Employee retention	2	62	4.5
Technology	0	40	4.3
Labour outsourcing	6	9	3.1
Seasonality	15	19	3.1
Employee turnover	27	4	2.2

* *on a scale from negative influence (1) to positive influence (5)*

When asked to comment on what they think would most help improve productivity in their business, managers had a wide range of suggestions which were very similar with those raised in the interviews. Staff attitude to work and ability to work autonomously was the most common factor suggested (22.5%). Staff retention/experience, staff skills/qualifications and training were each commented on by 10% of respondents. Financial considerations at a macro level (e.g. taxation) were raised by 15% of respondents.

Communication and networking was also considered important by 15% of respondents. Supply related factors were raised in a further 20% of cases.

Over half of national survey respondents (52.3%) said that the primary barriers to improving their productivity are finding and retaining good staff, lack of qualifications/knowledge/skills, and poor employee attitudes to work.

Difficulties with the property/product upon which the business is based were also a cause for some concern. Interestingly very few workplaces featured productivity incentives with the exception of some of the larger enterprises.

Some comments illustrated the frustrations experienced by some managers:

> *"In my experience, most people will do the least amount of work required until management are forced to make them be more accountable..." The majority of employees are not given individual productivity-related targets to meet, although non-monetary targets such as quality service are often emphasized. Employees in all focus groups said that general head count, 'table turns', smoothness of daily operations and how much they 'up sold' in a day are good indicators of productivity. Managers were often perceived as "discussing numbers behind closed doors" and targets are not communicated to employees, or the targets do not mean much to the employees.*

Those working in larger organizations are more likely to have financial targets to be met each day and/or each week; and they are also encouraged to up-sell by offering customers drinks with their meals and desserts in order to meet targets. In larger enterprises, there are individual staff incentives for achieving productivity related targets, such as prizes and rewards:

> *"If we sell an extra breakfast we get one point, if we upgrade a room we get two points. When we reach 400 points we can stay for one night in the hotel free. For 20 points we may be able to park our car for one day free".*

Employees said that being efficient at what one does is an indicator of productivity. This includes taking the orders correctly, preparing and serving food quickly, reducing food wastage, and ensuring there are enough resources (i.e. clean glasses and cutlery) to provide to the customers.

Participants across all focus groups said good customer service is an essential indicator of productivity. For them good customer service includes *"being really nice to the guests"*, *"being fast"* but *"not too fast, which can be seen as abrupt"*, having strong abilities for multitasking and *"doing little jobs, which usually don't belong to anyone in particular"*.

The Central Role of Management

The research presented above reveals quite clearly the central importance of effective management in achieving higher productivity in the F&B sector. In an industry as labour-intensive as F&B, people, and their skills, are central to all attempts to enhance productivity (D'Annunzio-Green, Maxwell & Watson, 2000). Research has shown the positive relationship between effective human resource management and superior organisational performance in many different sectors and nations (Drucker, 1994; Huselid, 1995; Wright and Haggerty, 2005; Neal, West & Patterson, 2005). Such an approach is consistent with current views on the positive role that empowerment and high-involvement work practices can play on productivity improvement.

Such approaches shift the focus from scientific management practices (i.e. task simplification and specialisation) to one that relies more on the intellectual contribution of employees, and emphasises the responsibility for productivity lying at the employee, rather than solely at the management, level. Gavin and Mason (2004) recommend that organisations be designed and managed to promote worker satisfaction, and suggest that healthy and happy employees are more productive. In other research service strategies that took account of the need for employee development and well-being were found to have a positive effect on employee satisfaction and productivity (Goldstein, 2003).

As the Department of Labour (2004) notes in its discussion of productivity drivers, there is a:

> *"need to make sure the activities that create value within a firm are aligned with each other and with the overall business strategy, and that they are functioning effectively."*

This can be aided by employee participation.

> *"It is critical that employees at all levels of a firm have an opportunity to contribute to work organisation and to provide relevant practical advice from their respective positions."*

It is commonly accepted that effective leadership will lead to higher organisational performance and poor leadership will result in poor performance (gaard et al., 2005; Simons 2002). In its discussion of productivity drivers the Department of Labour also notes that:

> *"Leadership by example plays a strong role in creating a positive and productive workplace culture, and inspiring others to pursue those opportunities which have been identified. Leadership depth is important."*

Across all the sectors represented in the interviews, most managers are aware of their role in enhancing productivity.

"I am the conductor, I can put my sticks up and down and others have to perform. My performance is determined on good service, standards being met and the financials being sound".

"No one likes to work for a bad manager. I have learnt from good managers I have had in the past and then applied the good things gained from them. I still watch good managers now and whenever I go into bars and restaurants I watch how their staff work. For a while I did this every month at a cafe in Auckland to learn and observe".

Two management types emerged from the interviews: non-adaptive and adaptive. The former refers to a leadership style underpinned by a more paternalistic philosophy. The adaptive approach, by way of contrast, refers to a leadership style that emphasises things like team tasks and targets, and enhancing employees' readiness to participate in meeting overarching requirements.

It is not easy to generalise which of these approaches is superior in terms of F&B service productivity. The value of each depends on the environment in which the business is operating and the organisational culture and structure it exhibits, along with a range of other variables.

Employee Autonomy and Communication

According to many of managers interviewed, autonomy is important for hospitality employees especially when it comes to the spontaneity involved in the interaction between staff and customers. However, a culture of "controlled autonomy" also exists in some cases-especially in the larger enterprises-and is designed to control the major productivity parameters such as service standards, wastage, human resource costs. Smaller businesses, constrained by heavy and unpredictable workloads and limited resources, tend to give their staff more freedom.

Most of the managers feel that their employees have autonomy in their jobs. The difference lies in the extent to which formal methods are used to enhance communication. The larger organisations-especially hotels-have more formal consultation systems such as 'tea and scone' sessions and regular meetings where staff and management can discuss ideas and performance. As one manager from a larger organisation noted:

"Decision-making in an organisation this size is slower. Roll out of information is slower. No getting away from it, so the key is being smarter about how they make and implement decisions."

Managers were asked in the web survey how they would characterise the level of input their employees have in suggesting how the workplace can be better organised? The majority of the 55 managers who responded feel that their employees have either reasonable (49%) or significant (29%) levels of input into workplace organisation; only 4% believe their employees have no or very limited input.

The ability of staff and management to communicate effectively is vital to the creation of lower-stress workplaces. Stress is associated in a number of studies with reduced work output and poor employee performance (Lo & Lamm, 2005). Although respondents in an Auckland hotel study did not report high levels of work-related stress or stress-induced illness (Lo & Lamm, 2005), other research points to the fact that unpleasant, and stressful conditions in hospitality workplaces will have a negative influence on service quality, employee productivity and business performance.

Most employees feel that their managers rarely if ever discuss productivity concepts with them. Employees of larger organisations said their relationship with their managers is often formal, with some managers being very authoritarian to the point where everything is done as dictated, and no ideas are to be expressed. Participants in the smaller organisations also came across authoritarian managers and were usually disappointed in terms of productivity outcomes:

> *"My manager does not really encourage us to give input she just wants us to do what she asks. We are not encouraged to be creative, just to follow exactly what we are told. Once I put some magazines out for customers but was told to remove them. After that I didn't use my initiative as she didn't want our ideas".*

Some employees, mostly from smaller enterprises, have managers and supervisors who are more open to staff ideas about improving productivity and the work environment:

> *"We can make suggestions to our manager and he will discuss it with people on the same shift. He is willing to try new ideas".*

Those employees in large organisations commented that being casual and/ or part time, they don't have much contact with their manager, and thus *"couldn't care less as long as I get paid"*. Quality relationships between staff and management clearly have a bearing on whether the employees feel there is a good climate in which to discuss productivity, and how encouraged they feel to share ideas.

Feedback from managers on employee achievements plays a significant role for some employees in creating job satisfaction. Regular appraisals, pay rises and tips are stated as additional useful monetary indicators of productivity:

> *"One of the main drivers for me to be more productive is recognition, like one of the managers saying thank you at the end of the day."*

Smaller businesses are very positive about seeking staff input, but usually in an informal way with the owners and managers being happy to have staff input or ask staff about how things are going, often through just chatting with employees.

> *"The internal communication among the staff, from me and the staff...keeps the whole thing flowing very well and very productively...."*

The workplace climate both in terms of manager encouragement, as well as other employee attitudes, does not always encourage individuals to improve their work practices, as is illustrated:

> *"From my experience it is more like tall poppy syndrome. If you do something outside the norm everybody else may get niggled because they don't want to push themselves as much. It can be frustrating".*

Participants also admitted that making suggestions sometimes caused unwanted arguments, and organisations (especially larger) often have set ways of doing things making it hard for suggestions to be implemented. Some participants expressed doubt if their suggestions ever *"make it to the head office"*.

Good communication in a team has a significant impact on the efficiency of staff and their ability to perform the *"dance"*, and helps to improve the customer experience. It is especially crucial, but hard to achieve, when there is *"not time to say please and thanks"*.

Employees in larger organisations said that communication between staff was better in smaller teams where everyone listened to each other to avoid making mistakes, which helped to reduce conflict at the workplace. These same employees explained how Christmas and social events they have experienced gave them a chance to socialise which enhanced workplace relationships and fostered greater teamwork.

On several occasions focus group participants mentioned that conflict between kitchen staff and wait staff has a significant impact on overall productivity. Wait staff often get caught in the middle when customers are unhappy with their meal and the kitchen does not appreciate the criticism.

Employees think that poor English greatly affects customer service and productivity, especially in the case of phone orders. Chefs with poor English can be a particular cause of conflict among staff, exacerbating the antagonism between front of house and kitchen staff.

"Customers often got annoyed and even 'grumpy' when staff speak poor English, which effectively reduces customer satisfaction".

Quality of Working Life

The majority of managers interviewed said that they feel their jobs could potentially offer a comfortable level of work-life balance, but most choose to work a large number of hours. Many spoke of their passion for what they do, rather than viewing it as work.

"It is full on, it is my fault though. I work 6 days, 60-70 hours a week. I do what I know, and I still really enjoy my job, especially the project work as we have so many resources".

While several managers commented about the fact that they do not demonstrate good work-life balance themselves, they stated that they are highly aware of the importance of it for their staff productivity.

"Work-life balance is very important-they [staff] are rostered off with a weekend off once a month, and are not allowed to come in. We try to roster the days they want off".

The managers who responded to the web survey were asked if they think hospitality is an attractive career. Of the 53 participants who responded, 75% stated that they did feel it is attractive, with only 25% disagreeing.

Employees were less sure about the ability to manage work-life balance in a F&B services role, and less certain about F&B offering attractive career options. Working too many shifts can be a "novelty" at first, but inevitably results in one being overworked and stressed out. Most employees felt it is harder for permanent employees to achieve work-life balance, making some employees frustrated and leading them to seek less demanding, more motivating, and better paid work.

Employees from a range of organisation types said that hospitality work is *"hard and tiring and poorly paid"*. Negative comments about hospitality work also referred to being paid only minimum wage, working long hours, not getting paid overtime, rarely getting tips and bonuses, stressful work environment, and having to deal with negative customers. Uncertainty over work-flows was often commented on:

"If we are short of staff everyone has to do anything. I have done room service, a la Carte, as well as cashier. When we are not busy it is a totally different pace".

The overwhelming majority of employees say they will not be working in hospitality in the future. Those studying at university plan to quit their hospitality jobs after completing their studies. Among the reasons for quitting are: it's not a *"real"* career; it's not challenging enough; it has negative customers, and low pay. Employee dissatisfaction with high levels of formality at the workplace, as well as their desire to change their job is also apparent in larger enterprises.

"Never took it to be a career, just the hours suited".

There were a number of participants from a range of (mostly small) organisations who genuinely do enjoy working in the industry. These individuals are more likely to be working in establishments where their contribution is sought and valued, and display the attitude and enthusiasm for the industry demonstrated by the managers interviewed. For some employees flexibility in their work environment was also a real positive:

"I have flexibility at work; they are not too strict as long as you do your job properly. The cafe is busy but there are times when there is not too much pressure".

Some employees in larger organisations also believe that hospitality is becoming more professional, offers more opportunities, and people are*"starting to get paid more"*.

Support Networks

As the Department of Labour (2004) notes:

"Firms do not operate in isolation and there are significant productivity gains to be achieved by improving the exchange of knowledge, information and ideas through both formal and informal networks".

Those employers that completed the web survey use a variety of sources to help build managerial capability. The web survey participants were asked where they went to for help as an employer/manager. The two most popular sources are friends and personal networks (73%), followed by accountant or lawyer (45%). The Department of Labour ranks third as a source for those seeking assistance (43%).

Productivity is a topic for discussion for some managers with colleagues in their internal and external networks. External parties were often identified as trustworthy sources with which to discuss productivity, particularly regarding new products and process innovations.

About a third of the managers said that they do not really discuss productivity, especially beyond the walls of their own establishment.

"Productivity is not talked about in NZ like it is overseas-more a cost balancing act here. The industry is not so good at discussing productivity."

For employees networks also play an important role. In many cases ethnic or friendship networks are important in the hiring process-with recommendations for existing staff playing a role in the attraction of new workers. Within firms the 'social capital' associated with close knit work teams was also often mentioned-with effective team work seen a great way to learn new skills and create a more enjoyable work environment.

Hiring & Training

According to Fair and Brook (2001), businesses should "never underestimate the contribution to productivity that ongoing everyday training makes". At the same time the Department of Labour (2004) stresses.

"Skills shortages choke off growth potential. The knowledge, ability and skills of workers contribute to workplace productivity. Ongoing investment in foundation, technical, supervisory and managerial skills, together with improvements in work organisation, can help improve the productivity and performance of New Zealand firms."

The link between training and productivity is reflected in recent comments by Bruce Robertson, CEO of the Hospitality Association of New Zealand.

"One of the other keys to improved productivity is training. Three trained staff will always outperform four untrained staff and that applies right throughout the industry".

Research has shown that effective hiring and training is linked to a better ability to meet the needs of a target market (Shaw & Patterson, 1995), greater organisation commitment (Roehl & Swerdlow, 1999), more effective use of new technology, and enhanced work outcomes.

A firm can only achieve higher productivity by employing the right individuals with the capacity and desire to contribute to its goals.

"Good hiring creates a team; good communication creates a feeling of involvement, that it's more than just a job..."

Managers were asked in the web survey to identify the key characteristics or skills that they look for when recruiting staff. The two most commonly mentioned characteristics are honesty/loyalty, and experience (30% each). Personality and ability are sought by 23% and 20% of managers respectively.

Increasingly, these managers are recruiting from international sources and new migrant groups-this can lead to language difficulties. Managers were asked in the web survey about the adequacy of numeracy, literacy and language skills among applicants. Over 80% of managers feel that overall, applicants do have adequate language and literacy skills, although only 66% test language and

59% test literacy skills when hiring. Only a third of managers test for numeracy skills when hiring.

Training can positively impact on employee productivity, promoting customer satisfaction and sustaining competitive advantage. Training is associated with improved job satisfaction and intention to stay, and with reduced workplace problems and staff turnover (Poulston, 2006).

Most managers interviewed for this study believe training has very positive and direct impacts on productivity improvement. Some rank it the number one factor and most of them describe it as having *'important'*, *'crucial'*, *'huge'* and *'massive'* importance. On-the-job, in-house, and self-training are the most common forms of training in the industry. Constrained by cost, many the interviewees from smaller enterprises expressed their inability and reluctance to employ outside trainers, although a couple of them talked about using outside trainers in specialist areas.

These findings are confirmed through the web survey which indicates that 88% of managers say the most common form of training engaged in is informal and on-the-job. This is followed by formal in-house training (40%) and training by an external provider (31%). Over half the managers think the training they are providing to their staff matches their needs. However, constrained by time, money and business pressure, a number of managers expressed their concern about the lack of training they were able to provide.

Tertiary/private training is considered to be too theoretical and as not meeting the industry's requirements well. Most of the managers interviewed, and 60% of web survey participants, feel formal tertiary/private hospitality qualifications do not adequately prepare people to enter employment in their workplace.

"I think on the job training is very important-it is hard to take a student from tech and put them in this environment".

A theme of *"wanting to train them our way"* emerged amongst some of the managers interviewed. Some managers expressed annoyance at needing to at times 'untrain' those that had been trained to do jobs one way.

Employees had generally undergone some type of training when they first started their job; and the majority had on-the-job training provided by other staff and/or the manager. During on-the-job training, participants appreciate the opportunity to see senior staff *"showing tricks of the trade"* in the context of the workplace, and to ask questions. Employees said training provided by the manager rather than senior colleagues was most valuable, as other staff often *"forgot little things"* or deliberately *"withheld information"* on the best ways to do things. Some didn't have either formal or on-the-job training and were forced to follow a *"sink or swim"* approach-*"You are given a till and a bag of money and told to go for it"*.

Employees in larger organisations tended to have taken formal induction courses. During induction training, participants watch videos, are given manuals and introduced to key procedures. Research shows that employees of small and medium enterprises are less likely to have the opportunity to take part in this type of training than those working in large organisations Milne et al. (2004).

Employees in the focus groups generally believed that *"training helps to access jobs"*, but were less certain about its ability to provide access directly to managerial level jobs. Those in small organisation focus groups felt that there is a *"gap between education, hospitality and employers' expectations"* and most expressed some scepticism about the ability of tertiary providers to really prepare people for the F&B workforce.

TURNOVER & RETENTION

As one F&B manager Interviewed Noted: "Employee retention is important for consistency of products and customer service standards, so customers know what to expect".

Hinkin and Tracey (2000) list the following productivity-related costs of staff turnover: pre-departure productivity loss; learning curve; errors and waste; supervisory and peer disruption. Turnover can also negatively affect productivity within interdependent worker groupings and teams. However not all staff turnover has negative implications.

Moderate turnover can help to replace unproductive employees, enable career progression, enhance job and employee matching and is therefore viewed as enhancing productivity by some (Ilmakunnas et al., 2005).

A recent study of Auckland hospitality workers (Poulston, 2006) indicated short lengths of employment are common (87% have been in their current job fewer than five years), and that the majority work part-time. The latest survey conducted by the Restaurant Association (March 2006) indicates an average annual staff turnover of 29%, up from 26% in 2005. A New Zealand study of productivity found that high-involvement work practices and employee retention were strongly associated with strong enterprise performance (Guthrie, 2001). While turnover increases exposure to disruptions associated with the loss of employees in high involvement work practices this damage is greater than in settings where processes are more automated, for example the preparation areas of fast food establishments. Several managers stated in the interviews that they have real problems with staff turnover, tending to attribute much of the problem to the tight labour market rather than to factors that are directly within their control. Of the managers who completed the web survey question asking whether staff turnover affects productivity in their business, 54% feel it does, 29% feel it does not, and 17% responded 'not applicable'. The need for training new staff was seen as having the greatest negative effect on productivity. The time and cost involved are also detrimental effects stemming from staff turnover.

"Turnover is costly and involves extra training-there's also a trust issue with long term staff".

Large hotels are the most proactive in developing strategies to cope with turnover and to increase retention. Such strategies include succession planning, valuing internal communication to develop teams, and offering incentives for staff to stay. Training and flexible hours are also seen to be important in improving retention.

Overall, employees agreed that new staff often reduce productivity, because they are inefficient in their job, and because it is a burden on experienced staff to train someone new. Some employees, however, give new staff credit saying they usually are the most willing to learn.

Longer term staff not only have more industry experience, which helps to increase overall productivity, but also a stronger relationship with the kitchen staff, seen by many as prerequisite for having a *"good night"*.

Technology

In its discussion of productivity drivers the Department of Labour has identified that:

> *"innovation is a key part of raising workplace productivity. The appropriate introduction of advanced technology is linked with higher productivity, greater market share and employment growth".*

The introduction of advanced technology has been cited by several authors as an important and positive influence on productivity. Positive impacts however do not simply 'appear', rather they stem from the competent use and exploitation of IT capabilities (Sigala, 2003).

Technology has been found to 'support employees, enhance the quality of service, improve efficiencies, gain competitive advantage, maintain relationships with customers and increase profitability', while poor investment in technology (along with poor training) is likely to reduce profitability.

Several studies have shown that SMEs in the hospitality sector have special difficulties in maximising the productivity potential of information technologies.

Such enterprises are more likely to recognise the limitations of technology and emphasise the importance of 'the human touch'. At the same time SME owners and managers are more likely to have to out-source aspects of IT maintenance and may feel more removed from its application to the business setting.

The web survey found that 89% of managers have introduced new technology in the last two years with the view to improving productivity in their workplace.

The interviews revealed a similar trend. A range of technologies have been introduced by managers, primarily in the front of house area with computers being the most common purchase. Smaller businesses have a slower uptake of 'cutting edge' ICT devices such as wireless personal digital assistants and text message ordering.

Both managers and employees are divided on the issue of technology and productivity. Some think it enhances staff efficiency, saves time and improves productivity. Others say it in fact creates inefficiencies when it breaks down or requires updating.

While technology was discussed as an important contributor to productivity, most of the technology identified as bringing real efficiency gains is in the front of house environment. In particular, technological innovations are central to streamlining ordering and point of sale processes.

FOOD HOLDING TEMPERATURES

Two and Four Hour Rule

Hot food must be kept at 63°C or above. It must either be consumed within 2 hrs from the time the food temperature drops below 63°C, or be disposed of.

Chilled food may be displayed (away from under temperature control) for a maximum period of 4 hours. Only one such period is allowed, no matter how short. If, for example a refrigerated dish has been on display for 1 hour in a non-refrigerated environment, refrigerating it does not mean it can be displayed for a further 3 hours later. Where food is being cooled prior to refrigeration, cooling is to be achieved as quickly as possible as described in paragraph. When hot or cold food is displayed outside a temperature controlled environment, it is to be consumed or discarded.

Target Temperatures

Food Safety legislation and the Dept of Health require that certain temperatures be maintained throughout the food storage and production process.

- Cook
- Temp
- Service
- Temperature
- (At or above)
- Reheat Reheat
- Scotland
- Chilled
- Storage
- Frozen
- Storage
- Legally Required
- Temperature
- 0°C for 2 mins 63°C 75°C 82°C 8°C-18C.

Temperature Measurement and Temperature Probe Calibration

Measurement. A digital thermometer with attached probe is to be used for measuring food temperatures, including final cooking temperatures. The probe is to be disinfected with food safe wipes between measurements. There are also 'Between Pack Probes' are available that are to be used to measure temperatures of frozen and chilled foodstuffs, primarily at the delivery stage. Details of these items can be found in JSP 308.

Numbers of parts as available as follows;

Description	*NSN*	*Remarks*
Wipe Clean Thermometer	7320 99 0173073	
Robust Penetration Probe	7320 99 2154600	
Thermometer Between Pack Probe	6685 99 9906711	6 x 200 pack
Bacterial Probe wipes	7320 99 995890	
Pocket Stick Thermometer	6685 99 0007159	

Calibration: The digital thermometer is to be re-calibrated annually or in accordance with the manufacturer's instructions and warranty. A monthly calibration check is to be undertaken with an appropriate device in accordance with the manufacturer's instructions, and recorded on the template at Annex C to this chapter. The Test Plug is also to be calibrated in accordance with manufacturer's instructions and warranty.

Records

Records. Records are an important aspect of any due diligence defence and are essential in proving that the temperature requirements were being met in the event of any prosecution or food poisoning outbreak. Records (including Calibration records) are to be retained for 6 months. The mandatory records that are to be kept with unit files are listed below:

a. Delivery monitoring records.

b. All refrigerator and freezer temperature monitoring documents.

c. Food temperature monitoring records.

d. Calibration records for temperature probes.

Personal Hygiene

Good personal hygiene is a legislative requirement, ensuring safe food. Induction training must ensure that new staff joining are aware of the required standard Food handlers are a potential source of bacterial and physical contamination of food, and so personal hygiene is a key element in ensuring that food is prepared safely.

Staff training is to include all the basic elements of personal hygiene (including hand washing training) covered in this section and personnel are to understand the relevance of the precautions.

The FS(GFH)R lays down criteria for the personal hygiene of food handlers and the actions to be taken if a food handler is infected. These regulations require every person working in a food handling area to maintain a high degree of personal cleanliness.

Any person suffering from a disease that is likely to be transmitted through food must inform the person in charge of the galley/kitchen. A separate regulation states that any such infected person may not be permitted to work in any food handling area, in any capacity, where they might contaminate food.

Specific Requirements

a. *Hand washing.* The hands of food handlers are the principal agents in the transference of bacteria to food. Hand washing facilities (a separate basin that is only to be used for hand washing) in galleys/kitchens and heads/toilets are to include hot and cold running water, soap, nail brush and a suitable means of drying the hands. Regular checks by senior staff are to be made to ensure these facilities are available and are being used effectively. As a minimum, hand washing, is to take place:

(1) On entering a food room.

(2) Between handling raw and cooked food.

(3) After handling waste food or refuse.

(4) After smoking or eating.

(5) After visiting the WC.

b. *Cuts.* Open cuts harbour bacteria and must be covered with clean blue waterproof dressing to aid detection. Stocks of these are to be available in food preparation areas and readily accessible to food handlers.

c. *Jewellery.* The wearing of wristwatches, ear rings ('Studs' are acceptable) and other exposed body piercing, bracelets and rings, is not acceptable as they harbour bacteria and there is a risk of physical contamination of food. The wearing of a plain wedding ring is acceptable.

d. *Smoking.* Smoking is not permitted in food areas, as it transfers bacteria from the mouth to hands. Cigarette ends and ash also pose physical contamination risks.

e. *Protective Clothing and Changing Facilities.* Food handlers must wear suitable clean protective clothing (including appropriate footwear and hats) to prevent contamination of food from normal clothing. Such clothing is to be changed at the end of a shift, or sooner if the situation requires, in order to maintain hygienic standards. It is not to be worn outside food areas and associated premises. If necessary, an outer garment is to be worn over protective clothing whilst away from food production environment. Adequate changing facilities must be provided, with locker space for clean and soiled protective clothing. Changing facilities must be kept clean and tidy at all times.

f. *Visitors.* All visitors to a food preparation area are to be viewed as potential sources of contamination. They must therefore be provided with protective clothing and briefed upon good food hygiene practices before entering the food area. Before entering a food preparation area, visitors must confirm that they are not suffering from diarrhoea and/or vomiting, or heavy cold. Individuals who are suffering should not be allowed to proceed.

g. *Eating/drinking.* Food handlers are not to eat or drink in food rooms. It is acceptable for cooks to taste dishes during preparation in a manner that does not contaminate the food i.e. a clean spoon each time.

h. *Toilet Facilities.* Where possible toilets for catering personnel are to be separate from those for non-catering personnel and visitors. Toilets are not to be located directly within a food preparation area. There is to be an intervening ventilated space between toilets and food rooms. Food is not to be stored in that space. Toilets must be ventilated such that associated odours are prevented from permeating into food rooms. They must also be kept clean and tidy, in good repair with adequate supplies of toilet paper, hand washing facilities, nail brushes, soap and hand towels.

INFECTED FOOD HANDLERS

Dispersed Feeding

Dispersed feeding is the production and transportation of food for consumption away from an established unit/galley/kitchen or mess for example to personnel on security duties who are unable to be fed in-mess, non-public section functions, such as barbecues or sports meals. This method of feeding is recognised as high risk, because the majority of foods used require controlled temperatures during transportation in order to comply with Food Safety Legislation. Dispersed feeding should only take place when there is no suitable alternative.

Documentation. Before food leaves the kitchen, the Shift NCO (or equivalent if DEL or contract catering staff) is to ensure the completion (in duplicate) of FCAT 1013.

Dispersed Feeding Record (demanded from DSDC Llangennech)

A copy is to be retained with the food safety records and a copy handed to the person (recipient) taking the food out of the kitchen. The recipient is to be briefed on the following points:

a. Chilled Products. Chilled foods are to be consumed within 4 hours of being removed once cold products leave from a temperature controlled environment i.e. refrigerator or chilled servery and when and the product temperature rises above 8oC, products are to be consumed within 4 hours.

b. Hot Products. Hot foods are to be consumed within 2 hours of being placed into a container. Food is to be served at 63°C. If a hot container meal arrives at its final destination and the product temperature is below 63°C, it is to be consumed as soon as possible (but always within 2 hours from when the product was placed into the container). This time will have been recorded in column (b) of the FCAT 1013.

c. When temperature monitoring is unavailable at a dispersed feeding location, the food is to be consumed within 2 hrs (hot product) or 4 hours (cold product) of being placed in the insulated container or leaving a refrigerated environment. Once again the time of service is to be recorded in column (b) of FCAT 1013.

Transport Of Food

Movement of temperature-controlled foods may include:

a. The collection of food from suppliers and transportation to the main ration stores and messes.

b. The transportation of prepared food from messes to satellite or dispersed feeding facilities.

Transport can include items such as trolleys, bags, boxes, trays and crates. These articles must be kept clean. Where transport is also used for materials other than food, it must be thoroughly cleaned between loads to avoid the risk of cross contamination.

Non-catering personnel collecting foodstuffs from the mess are to be briefed on the importance of maintaining the integrity of the food chain, and how this is achieved.

During transport, all food is to be covered and is to be transported in appropriate insulated containers. High-risk food must be kept separate from anything that could cause contamination. When transporting prepared meals, dispatch and receipt temperature checks are to be recorded in accordance with paragraph.

Cleaning

The FS(GFH)R requires adequate facilities are provided for the cleaning and disinfecting of work utensils and equipment. These facilities must be constructed of materials resistant to corrosion, must be easy to clean and have an adequate supply of hot and cold water.

Equipment items that come into contact with high-risk foods will need to be cleaned and disinfected as necessary. Facilities must be provided to clean and disinfect all tools and equipment, crockery, cutlery, glasses and serving dishes. There are to be sufficient cleaning materials and equipment to match the size of the catering facility. Staff involved in cleaning must wear suitable and adequate protective clothing, which must be changed, at the end of each shift.

Colour coded cleaning equipment, e.g. mops must be cleaned, dried and stored in a designated area once finished with. Drying of equipment must not cause recontamination, e.g. from soiled cloths. A full list of NSNs of colour-coded equipment is shown in JSP 308.

Equipment must be of durable construction and resistant to corrosion, especially items that will come into contact with powerful cleaning chemicals.

Sinks. It is a legal requirement that, where appropriate, adequate provision must be made for any necessary washing of food. Every sink or other facility for the washing of food must have an adequate supply of hot and/or cold potable water as required, and be kept clean. Separate facilities must be provided for hand washing, food preparation and equipment washing if the volume of preparation in the kitchen demands it. In small catering operations, one sink may be used for both equipment and food washing, provided that both activities can be done effectively and without prejudice to food safety, i.e. the sink is to be thoroughly cleaned between each process. It is good practice to have signs above sinks indicating what they can be used for. Hot water supply is not essential if a sink is to be used exclusively for food preparation.

Disinfection. Disinfection is a process that reduces microorganisms to a level that will not lead to harmful contamination or spoilage of food. Sterilisation is a process that destroys all living organisms. The word 'sterilise' is sometimes used in food hygiene literature and in some official publications where the word 'disinfect' would be correct. Cleaning products are often described as 'detergent/sterilise' when 'detergent/disinfectant' would be a more accurate description. The following definitions clarify the relationship between disinfection and cleaning:

a. A disinfectant is a chemical agent used for disinfection after cleaning.
b. Sanitisation is a term used mainly in the food and catering industry. It is a process of both cleaning and disinfecting utensils and equipment.
c. A sanitiser is a chemical agent used for sanitisation.
d. Sterilisation is a process intended to destroy or remove all living organisms. It is not to be confused with disinfection.
e. Cleaning, Heat, and Chemical Disinfectants. Disinfection is essential in food hygiene. It is achieved by:
   (1) Cleaning. This is the most useful and practical method of removing food residues, dirt, grease and other undesirable debris. It can be used alone or in combination with the other two methods. A high proportion of microbial contamination is removed from equipment, food preparation surfaces and hands by thorough cleaning.
   (2) Disinfection by Heat. This is considered the most reliable of the three methods of disinfection. Immersion in water at 65ºC for 10 minutes, or at higher temperatures for shorter times can be relied upon to destroy most microorganisms harmful to health, with the

exception of bacterial spores. Boiling water is an effective and cheap disinfectant although bacterial spores may survive boiling. The temperature of the washing or rinse cycle in a dishwashing machine can be adjusted to include heat disinfection in the cycle, in accordance with the manufacturer's instruction.

(3) Some chemical disinfectants are active against a wide range of bacteria; others have a narrow range of antibacterial activity. In all circumstances chemical disinfectants must be used in accordance with the manufacturer's instruction. It is military policy to use general-purpose detergent for general cleaning with surface sanitisers. Chemical disinfectants (stored in a designated

Cleaning Procedures

Cleaning of food premises is carried out to remove dirt and grease from all surfaces and equipment, including pathogenic bacteria, that could compromise food safety, as well as food debris. It is imperative that high standards of cleaning are maintained for the following reasons:

a. To comply with the legal requirements of the FS(GFH)R.
b. To reduce the risk of food poisoning by removing food residues, which could contain harmful bacteria.
c. To deny pests harbourage and food.
d. To reduce the risk of foreign objects physically contaminating food.
e. To promote hygiene awareness amongst catering personnel.
f. To provide a pleasant working environment and a favourable image to personnel.

Cleaning Schedule: A planned cleaning schedule programme is important in ensuring that high standards of cleanliness in all food areas are achieved and maintained. This can be achieved by adhering to a written cleaning schedule. Food preparation surfaces must be regularly cleaned whilst in use. This is particularly important where space is limited and the cross contamination risk is increased.

a. Instructions for cleaning specific areas or pieces of equipment are to include the following information relative to the task:
   (1) Job description.
   (2) Cleaning materials and chemicals to be used (in accordance with manufacturer's instruction).
   (3) Safety precautions.
   (4) Job method.

b. The cleaning schedule is to state how often a specific area or piece of equipment is to be cleaned and who is responsible for checking that all cleaning tasks have been completed to a satisfactory standard.

c. General information concerning the cleaning of food equipment, surfaces and the material structures, including a list of cleaning equipment and agents used in food rooms/kitchen/galley/mess, are at Annex D to this chapter.

d. Catering managers are to implement a cleaning schedule that relates to all food areas within their department. A schedule detailing the frequency of routine cleaning tasks should be produced for each catering facility. It is to be contained within Catering Standing Orders and communicated to all personnel (e.g. by displaying the schedule on a notice board). A matrix showing the frequency, details of the task, type of cleaning required and a signature block is considered the most useful layout and should be readily available for operatives to follow and sign off. A signature block should also be included for a supervisory check.

Deep Cleaning: All galleys/kitchens and associated areas are to be deep cleaned in accordance with single Service instructions. Catering managers are to ensure that deep cleaning contracts are adequate to meet the tasks required.

Food Preparation Areas

Surfaces. The FS(GFH)R require that those surfaces which come into contact with food must be maintained in a sound condition, be easy to clean and, where necessary, disinfect. This requires the use of smooth, washable and nontoxic materials. To comply with this legal requirement, it is important that surfaces that come into contact with high-risk foods must be capable of being disinfected regularly. Examples of such surfaces, assuming that they are properly fixed, applied or installed and maintained, include: stainless steel, ceramics, and food grade plastics. Wooden boards are inappropriate for the preparation of high-risk foods. Food preparation surfaces must either be continuous in their construction or have properly sealed joints.

Design and Layout. The design and layout of food areas are to:

a. Permit adequate cleaning and/or disinfection.

b. Protect against dirt, contact with toxic materials, and formation of condensation and mould.

c. Encourage good food hygiene practices and prevent cross contamination by foodstuffs, equipment, materials, water, air supply or personnel and external sources of contamination such as pests.

d. Where practicable, suitable temperature conditions should be provided for the hygienic preparation of food.

Structural Requirements. In areas where food is prepared, consideration is to be given to floor and wall surfaces, ceilings and overhead fixtures, windows, doors and other openings. There are to be adequate washbasins, sanitary conveniences, ventilation, and drainage. Any maintenance requirements and requests to catering premises are to be recorded in either the Maintenance of Catering Premises – Record of Work Services Template at Annex E to this chapter or the unit works service record. Priority awarded to work service request is entirely dependent on the type of establishment and the impact of the equipment failure. The following priorities are to be applied:

MOD Standards. Defence Estates (DE) Design and Maintenance Guide 18 deals with the design of kitchens and serveries for all ranks, and dining rooms for Junior Ranks. Details of dining rooms for Officers and SNCOs are given in JSP 315 Services Accommodation Code, Scales 29 and 34. The guide is also to be read in conjunction with DE Specification 42 on Catering Equipment and JSP 315, Scales 01, 39, 40, 45, 47 and 52. For HM Ships and Submarines, the relevant Naval Engineering Standards (NES) publication provides guidance in the design, layout and equipping of Galleys, Storerooms (refrigerated and dry) and associated areas.

PEST CONTROL

Pests are known to carry a number of pathogenic organisms that can be transmitted to humans through contaminated food. In addition, pests will damage food stocks causing financial loss. It is therefore important that food premises are kept pest free. The Prevention of Damage by Pests Act 1949 and the FS(GFH)R impose legal duties on owners and occupiers of buildings to keep their premises free from infestation.

A wide variety of insect and rodent pests will enter food premises for a number of reasons:

a. Food. Even in small quantities, food will enable pests to survive and multiply. Regular and thorough cleaning of spillages is therefore imperative.

b. Warmth. Pests of all types are attracted to buildings, which offer even limited warmth away from outdoor conditions. A few degrees increase in temperature will provide conditions in which breeding is enhanced and proliferation encouraged.

c. Shelter. Almost every building provides a variety of harbourages for pests.

Contrary to common belief, it is the newer buildings with suspended ceilings, panelled walls, service ducts and enclosed electrical trunking, which are more likely to create a problem, than older buildings without such features. Access must be provided to these spaces for the effective control of pests.

Denial of food, warmth or shelter will prevent the survival of pests. This form of control can be termed 'environmental control' and is the first line of defence against possible infestation. Environmental control may be considered as denial of access (proofing), food and harbourage. Flying Insect Control. Emphasis is to be placed on the environmental and physical control detailed previously to reduce the risk of food contamination. Areas around food premises are to be kept clean and tidy to reduce the number of possible breeding sites.

External refuse containers are to be clean and in good repair, and have tight fitting lids. If skips are used they are to be completely enclosed. Waste food containers must be washed out before being stored outside.

Windows and other openings which provide ventilation are to be fitted with close fitting and cleanable fly screens where required. Doors are to be kept closed or fitted with screens or clear plastic heavy-duty strips. Electronic fly killers, where required, are to be installed but must not be sited above food preparation areas. The bulbs within electric fly killers must be changed in accordance with the manufacturer's instructions.

Recording Sightings

All personnel employed in a kitchen/galley have a duty to report to the Head Chef any evidence of pest infestation. Kitchen Managers are to record such sightings in the Pest Management Register at Annex F to this chapter. Under operational conditions or on exercise, trained Service personnel carry out pest and vector or vermin control. In peacetime (except in some overseas Commands and HM Naval Bases where MOD Civilian pest control operators are employed) civilian contractors normally carry out pest control. There are two ways in which a pest control contractor may be used:

a. To deal with and eradicate a single infestation.

b. To act as a long-term contractor who will visit the premises regularly and carry out pest control treatments as necessary. This proactive approach is more suitable and is recommended.

Waste Disposal

The FS(GFH)R require that food waste and other refuse must not be allowed to accumulate in food rooms, except so far as is unavoidable for the proper functioning of the catering department. It is recommended that systems of work are in place to ensure that refuse containers in food rooms are not over filled and are emptied regularly. All waste is to be removed at the end of the working day. Food waste and other refuse must be deposited in sealed containers which must be of an appropriate construction, kept in sound condition, and, where necessary, be easy to clean and disinfect.

Lids are not required on refuse containers used for temporary storage of waste in food preparation areas. They are frequently touched by the hands of food handlers and may be a serious source of contamination. Any refuse containers used for storage of waste awaiting collection and removal are to have a lid and to be constructed of durable material, which makes them easy to clean and disinfect. Refuse containers are to be included in the cleaning schedule.

The FS(GFH)R state that adequate provision must be made for the removal and storage of food waste and other refuse. Refuse stores must be designed and managed in such a way as to enable them to be kept clean, prevent access by pests, and protect against contamination of food, drinking water, equipment or premises.

Areas for indoor storage of refuse must be remote from food rooms and not sited near the main delivery entrance. In establishments it is good practice to have a separate area designated for the storage of outdoor waste with well-lit hard standing. A hose is to be provided for cleaning purposes.

Environmental Protection (EP) and Marine Pollution Regulations (MARPOL)

Disposal of food waste must meet the requirements of EP (JSP 418), DEFRA instructions and MARPOL legislation. Revised legislation relating to the disposal of raw meat and fish produced as waste product came into effect 01 Jul 03.

Such products can no longer be sent to landfill for disposal. A transitional period until Dec05 has been granted before which DCG will publish policy regarding disposal of such products.

THE FIVE PRINCIPLES OF HAZARD ANALYSIS

Analysis of the Potential Food Hazards in a Food Business Operation

Food Hazard-A hazard is anything that could cause harm to the consumer. There are three main hazards that may arise with food served in catering premises. These are contamination of food by:

- Bacteria or other microorganisms that cause food poisoning.
- Chemicals, for example by cleaning materials or pest baits.
- Foreign materials such as glass, metal, plastic and so on.

Of these, the most important hazard is likely to be harmful bacteria or other germs that may contaminate and multiply in food. Identification of the Points in those Operations where Food Hazards may occur. Food passes through many steps; delivery, storage, preparation, cooking, cooling and so on. Hazards can occur at many or all of the steps. For each type of food, the hazards that may occur at each step are to be identified. Mostly these will be steps in the operation where:

- Food can become Contaminated with microorganisms, chemicals or foreign material.
- Bacteria can Multiply if the food is held too long at the wrong temperature.
- Micro organisms Survive a process that should kill them, e.g. when cooking or disinfection of equipment is inadequate.

Deciding which of the critical points identified are critical to ensuring food safety.

Critical Points are steps at which the hazards must be controlled to ensure that a hazard is eliminated or reduced to a safe level.

- Any step where food may be become Contaminated is to be controlled. Controls include clean and disinfected equipment, the personal hygiene of staff, and separation of raw and cooked food. All food is to be protected from contamination by foreign bodies, pests or chemicals.
- Steps where bacteria may be able to Multiply in food must be controlled. The time and temperature at which food is held, stored or displayed are likely to be critical.
- Any cooking or reheating step is to be able to Kill harmful microorganisms. It will be critical that heating is thorough. Cooking is normally the most important control step in most food preparation. Chemical disinfection of equipment is another control point designed to Kill microorganisms.

Identification and implementation of effective control and monitoring procedures at those critical points

Controls must be set for the Critical Points, then Checks Introduced

Controls will either reduce the hazard to an acceptable level or get rid of it completely. The controls are to be precise as possible. For example, it is better to state that certain products will be stored under refrigeration at a set temperature, rather than to simply say that it must be kept in the chiller. A control target is to be set for every critical control point that has been identified. When controls have been set, it is then possible to Monitor the critical points whenever that preparation step is used. The targets can be checked. The frequency of checks is to be set for each control. Checking temperatures does not always involve probing food with a thermometer. Delivery vans or storage chillers may be fitted with temperature monitors and these can be checked. Other critical controls are more difficult to measure, e.g. cleaning & disinfection of equipment or the personal hygiene of staff. They will be often vital to the safety of food, and there are to be regular checks that standards are kept up. Review of the analysis of food hazards, the critical control points and the control and monitoring procedures periodically, and whenever the food business's operations change.

It is not satisfactory to go through this process once and forget about it. Occasionally the system may need to be reviewed and amended, for example:

- If controls or methods of checking are found to be ineffective or impracticable.
- The menu changes. New dishes may have new hazards and controls.
- The method of preparation changes.
- New equipment is introduced.

ANNEX B-HAZARD ANALYSIS

The Four-line Method

This annex contains:

a. Flow Diagram-An Overview of the Four-Lines.
b. Flow Diagram – LINE A – Food/dishes served cold.
c. Flow Diagram – LINE B – Food/dishes served hot.
d. Flow Diagram – LINE C – Food/dishes cooked, chilled and served cold.
e. Flow Diagram – LINE D – Food/dishes cooked. Chilled and reheated.
f. Hazard Analysis Flow Chart – LINE A.
g. Hazard Analysis Flow Chart – LINE B.
h. Hazard Analysis Flow Chart – LINE C.
i. Hazard Analysis Flow Chart – LINE D.
j. Galley/Kitchen/Mess Management Records:
   (1) Daily Food Safety management Record (Template)

(2) Advance Food Preparation Record (Template)

(3) Food Time/Temperature Record (Template)

(4) Galley/Kitchen/Mess Fridge/Freezer/Blast Chiller Daily Monitoring Record (Template)

(5) Galley/Kitchen/Mess Fridge/Freezer/Blast Chiller Monitoring Record (Template).

Use of the Four-Line Method

All caterers are to use this guide as a reference document. Each catering establishment will have unique hazards, which are to be annotated onto the template as a Control Point, a Critical Point or a Critical Control Point.

The Four-Line Method is to be used as follows:

a. Monitoring and Recording of the HAL's. Record the daily menu on single Service record forms. Determine high-risk menu items and, by referring to the Four-Line method guidance, annotate on the record the relevant Hazard Analysis Line (HAL) to the individual food dish.

b. Whilst the Four-Line method provides a system of HAL within kitchen and provision areas, other factors will need to be considered in ancillary kitchen areas e.g. delivery bays and pot/crock wash.

FOOD HYGIENE IN CATERING ESTABLISHMENTS

Legislation and Model Regulations

Food and Agriculture Organization of the United Nations

The World Health Organization, through its Food Safety Programme, which is aimed at ensuring the safety of foods for the consumer, initiated a survey of legislation related to food hygiene in catering establishments. The results of the survey, which are summarized in this publication, and the model food hygiene regulations for catering establishments, also included here, were reviewed by a joint FAO/WHO expert consultation held in Geneva from 28 April to 2 May 1975.

This meeting was funded by the United Nations Environment Programme (UNEP) as part of its activity to "support, and accelerate or expand the work of the FAO/WHO Codex Alimentarius Commission on international standards for pollutants in food and strengthening of FAO/WHO capabilities to assist developing countries in food control". In this activity the Food and Agriculture Organization of the United Nations and the World Health Organization are cooperating agencies.

Many of the legislative texts on which this survey was based have been published, in full or in summary, in:

(1) International digest of health legislation. Geneva, World Health Organization (published quarterly); and

(2) Food and agricultural legislation. Rome, Food and Agriculture Organization of the United Nations (published quarterly).

The survey covers a reasonably representative cross-section of the legislation of Member.

States in all Parts of the World

Food control has two principal objectives: protecting consumers against health risks, fraud, and adulteration of foods and assuring fair practices in the food trade. The survey described here was concerned with the first of these objectives, namely, the safety and cleanliness of food, i.e., food hygiene.

The need for a high standard of food hygiene and for adequate control measures is particularly important in catering^ establishments of all kinds. The fourth report of the WHO Expert Committee on Environmental Sanitation-' points out that:

> *"One careless food handler, or one human carrier of disease, preparing food at home will jeopardize the health of only a small number of persons, mainly members of the family. When one such person works in the kitchen of a restaurant, hospital, factory, canteen, school, or other place where meals are supplied to many people, the number of potential victims is correspondingly greater."*

The present report is therefore concerned essentially with the promotion of food hygiene in catering establishments.

Nevertheless, since the basic principles of food hygiene are common to all types of food establishments and food handling activities, the model regulations annexed to this report can be used as the basis for the legal control of all types of food "Food hygiene" has been defined as, "The measures whereby the wholesomeness, soundness and safety for human consumption are secured or increased, covering all facets of food production, harvesting, processing, distribution, preparation and service and of possible causes of toxicity (physical, chemical or microbiological)".

Geneva, World Health Organization

The model code sets out in detail, but in simple terms, the various points which should be legally enforceable in any country and should be embodied in regulations or secondary legislation. It is intended, in particular, for the guidance of developing countries wishing to build up an effective organization for the control of standards of hygiene in catering establishments. In drawing up this code account has been taken of a statement by the Codex Alimentarius Commission on general principles of food hygiene.

Measures for Promoting Food Hygiene

Effective legislation is essential in any adequate system for promoting food hygiene, but this is only part of such a system. However comprehensive the legislation, it will be ineffective unless it can be enforced. This means that there must be an agency, or agencies, charged with that responsibility. These responsible bodies must in turn employ an adequate number of capable and well trained staff for enforcement duties. Indeed, one of the first steps which developing countries wanting to build up a service to improve standards of food hygiene should take is the development of a trained field staff.

Field staff cannot, however, be engaged solely in law enforcement activities. Their principal role must be to exercise constant surveillance over catering establishments so that they are fully aware of the standards of hygiene being observed and are able to take prompt action to correct any faults that they discover.

This work will involve frequent inspections of premises and the collection of samples for bacteriological and chemical analysis, and supporting services such as laboratory facilities will be required. The field staff will also have advisory and educational functions, obtaining improvements in the handling of food through persuasion and consent and the education of managements.

To establish and maintain an adequate field staff with the support services required and to meet related expenditure on, for example, transport, equipment and materials, will require adequate financial resources. Part of the finances required may come from licensing or registration fees or other charges on the catering industry, but the major part will usually have to be obtained from central or local governments, or both. Governments have to decide how the funds are to be provided but it may be necessary to present strong arguments to justify the request, particularly where a country has previously not had a comprehensive food control service. Since the primary object of a food hygiene programme is to protect the health of the public, it is not always easy to present a case on a cost-benefit basis.

It must be recognized that promotion of food hygiene is only one of a number of measures for protecting and improving human health that fall within the scope of environmental health. According to a WHO Expert Committee on National Environmental Health Programmes: Their Planning, Organization, and Administration, environmental health "refers to the ecological balance that must exist between man and his environment in order to ensure his well-being".

Within the field of environmental health, environmental sanitation was defined by a WHO Expert Committee in 1950 as "the control of those factors in man's environment which exercise or may exercise a deleterious effect on his physical, mental, or social well-being".

The report of Joint FAO/WHO Food Standards Programme, Codex Alimentarius Commission. Recommended international code of practice: general principles of food hygiene. Rome, 1969.

WHO Technical Report Series, No. 439, 1970.

Committee on Environmental Sanitation). the WHO Expert Committee on National Environmental Health Programmes lists 17 item considered to be included in, or related to, environmental health. Among those items are water supplies, wastes disposal, vector control, and food hygiene.

Food hygiene, therefore, is an integral part of a health service and it is important to appreciate that it is difficult to develop in isolation a satisfactory programme for raising standards of food hygiene. High priority should be given to achieving proper standards in this particular field, but progress will be made more easily if efforts are also made to improve general environmental health.

In other words, a food hygiene programme should be developed within the context of progressive environmental health and personal health programmes. Good food hygiene is impossible without adequate supplies of safe water, proper means for wastes disposal, effective pest control, and reasonable living conditions. It is also difficult to reduce the incidence of food-borne infections where disease is widespread among the population and adequate steps are not taken to reduce sources of infection.

5

Agriculture and Food Biotechnology

INTRODUCTION

Biotechnology is necessary to maintain our agriculture competitive and remunerative and to achieve nutrition security in the face of major challenges such as declining per capita availability of arable land; lower productivity of crops, livestock and fisheries, heavy production losses due to biotic (insects pests, weeds) and abiotic (salinity, drought, alkalinity) stresses; heavy post-harvest crop damage and declining availability of water as an agricultural input. Investment in agricultural related biotechnology has resulted in significantly enhanced R&D capability and institutional building over the years. However, progress has been rather slow in converting the research leads into usable products.

Uncertainties regarding IPR management and regulatory requirements, poor understanding of risk assessment and lack of effective management and commercialization strategies have been significant impediments. India owns very few genes of applied value. The majority of the genes under use-about 40-are currently held by MNCs and have been received under material transfer agreements for R&D purpose without clarity on the potential for commercialization.

The spectrum of biotechnology application in agriculture is very wide and includes generation of improved crops, animals, plants of agro forestry importance; microbes; use of molecular markers to tag genes of interest; accelerating of breeding through marker-assisted selection; fingerprinting of cultivars, land raises, germplasm stocks; DNA based diagnostics for pests/ pathogens of crops, farm animals and fish; assessment and monitoring of bio diversity; in vitro mass multiplication of elite planting material; embryo transfer technology for animal breeding; food and feed biotechnology. Plants and animals are being used for the production of therapeutically or industrially useful products, the emphasis being on improving efficiency and lowering the cost of production. However, emphasis should not be on edible vaccines for which use in real life condition is difficult. Nutrition and balanced diet are emerging to be important health promotional strategies. Biotechnology has a critical role in developing and processing value added products of enhanced nutritive quality and providing tools for ensuring and monitoring food quality and safety.

It has been estimated that if Biofertilizers were used to substitute only 25% of chemical fertilizers on just 50% of India's crops the potential would be 2,35,000 MT. Today about 13,000 MT of Biofertilizers are used-only 0.36% of the total fertilizer use. The projected production target by 2011 is roughly around 50,000 MT. Biopesticides have fared slightly better with 2.5% share of the total pesticide market of 2700 crores and an annual growth rate of 10-15 %. In spite of the obvious advantages, several constraints have limited their wider usage such as products of inconsistent quality, short shelf life, sensitivity to drought, temperature, and agronomic conditions.

From a research perspective the spectrum of organisms studied has been rather narrow and testing has been on limited scale and restricted mainly to agronomic parameters. Environmental factors such as survival in the Rhizosphere/phyllosphere and competition of native microbes have not received sufficient attention. Moreover, results on crops are slow to show. Unless there is a policy initiative at the centre and the state to actively promote Biofertilizers and biopesticides at a faster pace, there is unlikely to be a quantum jump in their consumption.

A taskforce headed by Dr MS Swaminthan (2004) under the Ministry of Agriculture has prepared a detailed framework on the application of biotechnology in agriculture. The report rightly lays emphasis on the judicious use of biotechnologies for the economic well being of farm families, food security of the nation, health security of the consumer, protection of the environment, and security of national and international trade in farm commodities.

GUIDING PRINCIPLES

Consistent with the overall vision outlined by MS Swaminathan taskforce, the priorities in agri-biotech would be based on social, economic, ecological, ethical, and gender equity issues. The following guiding principles would apply across the sector:

- A comprehensive and integrated view should be developed of r-DNA and non r-DNA based applications of biotechnology with other technological components required for agriculture as a whole
- Use of conventional biotechnologies (e.g. biofertilizers, biopesticides, bioremediation technologies, molecular assisted grading, plant tissue culture etc.) should continue to be encouraged and supported. A precautionary, yet promotional approach should be adopted in employing transgenic R&D activities based on technological feasibility, socio-economic considerations and promotion of trade.
- Public funding should be avoided to research areas of low priority or those that could reduce employment and impinge the livelihood of rural families.
- Regulatory requirement in compliance with Cartagena Protocol, another international treaty and protocol for biosafety, germplasm exchange and access and the guiding principles of codex alimentarius will be implemented through inter ministerial consultative process
- Transgenic plants should not be commercialized in crops/commodities where our international trade may be affected. However, their use may be allowed for generation of proof of principle, strictly for R&D, their alternate systems are not available or not suitable.
- In a long term perspective basic research for development of low volume, high value secondary and tertiary products through enabling technologies of genomics, proteomics, engineering of metabolic pathways, RNAi, host pathogen interaction and others. Research and support of biosafety regulation would need support.

It is proposed to do away with the large-scale field-testing of the released transgenic events and make it compliant to agronomic test requirements.

Strategic Actions:

(i) Accelerating the pace of product development

- In our quest for better products, strong and sustained support should be given to encourage indigenous discovery of new genes and promoters in both public and privately owned institutions. Nevertheless, wherever there be an urgent need for a product to achieve food or nutritional security, creative commercial and academic international partnerships should be explored in national interest for sourcing important genes and promoters through licensing arrangements on exclusive/non-exclusive basis. The cost effectiveness should be carefully assessed on a case-to-case basis.

- A gene bank should be created and be accessible to private and public sector organizations after payment of an appropriate fee.

(ii) Public-Private Partnership

- There is an urgent need to promote and improve the levels of horizontal integration between public-public and public-private laboratories.
- Institutions that generate knowledge and those that specialize in late stage field trials are currently compartmentalized. While support to public-funded innovation must continue to be strengthened, it is proposed that at least 30% of government-funded programmes must have a commercial partner who will be responsible for directing R&D towards commercialization. Public investment should also be encouraged in small and medium companies, especially for late stage trials of transgenic crops. Partnership between public-funded organizations and industry is crucial in the science-to-product chain.

(iii) Inter-ministerial Agriculture Biotechnology Board

- An inter-ministerial Agriculture Biotechnology Board involving Ministry of Agriculture, ICAR, DBT, MoEF, regulatory authority, expert scientists, industry, and the farming community should be established to continuously assess cross cutting issues such as: duplication of R&D investments; capacity building; promotion of horizontal partnerships between various components in the knowledge-product chain; the most cost-effective manner of overcoming nutrition deficiencies (viz. iron, zinc, iodine, vitamin A); availability, access, release and efficient system for biosafety assessment of GMOs and products thereof; safe use of approved technologies and prevention of unauthorized ones; building public trust and understanding biotechnological application relating to global warming, climate change and sea level rise; global trends in consumer/industry preferences of farm commodities. This will also monitor trade and collect market intelligence with respect to GM crops and products and follow the trend of organic markets and watch international developments to identify niche markets, monitor countries that are rejecting GM foods and feed this intelligence to concerned agencies.

PRIORITIES

Priorities for crops and traits should be set after conducting a need assessment exercise in various farming zones. However, an indicative list has been suggested by MS Swaminathan Task Force (2004).

Crop: Priority target traits in crop plants would be yield increase, pest and disease resistance, abiotic stress tolerance, enhanced quality, and shelf life, engineering male sterility and development of apomixis. Crops of priority should be rice, wheat, maize, sorghum, pigeon pea, chickpea, moong bean, groundnut, mustard, soybean, cotton, sugarcane, potato, tomato, cole crops, banana, papayas and citrus. In priority crops equal emphasis should be given to GM hybrids and new varieties. The varieties in contrast to hybrids, are preferred by small farmers as they can use their own farm saved seeds for at least three or four years. In case of hybrids, research on the introduction of genetic factors for apomixis would be supported so that resource-poor farmers can derive benefits from hybrid vigour without having to buy expensive seeds every cropping season.

Livestock: Priority target traits in livestock would be enhanced fertility and reproductive performance, improved quality, resistance to diseases for reduced drug use, production of therapeutically useful products and quality feed. Livestock of priority would be buffalo, cattle, sheep and goat. Emphasis would be given to animal healthcare, nutrition, development of transgenics and genomics. It is proposed to set up an autonomous institution for animal biotechnology.

Aquaculture and Marine Biotechnology: Application of biotechnology would be crucial in disease resistance, enhanced productivity, fertility and reproductive growth, use of aquatic species as bioreactors for production of industrial products, value added products from sea weeds and other marine taxa and biosensors for pollution monitoring. Species of priority in fisheries would be carps, tiger shrimps and fresh water prawns. It is proposed to set up under the auspices of DBT and autonomous centre for marine biotechnology

Food and Nutrition: R&D would be focused on: development of biotechnology tools for evaluating food safety, development of rapid diagnostic kits for detection of various food borne pathogens; development of analogical methods for detection of genetically modified foods and products derived there from; development of nutraceuticals/health food supplements/functional foods for holistic health; development of pre-cooked, ready-to-eat, nutritionally fortified food for school going children; development of suitable pro-biotics for therapeutic purposes and development of bio food additives. It is proposed to set up (under the auspices of Department of Biotechnology) an autonomous institute for nutritional biology and food biotechnology (2006).

Biofertilizers and Biopesticides: Priorities would include screening of elite strains of micros-organisms and/or productions of super-strains, better understanding of the dynamics of symbiotic nitrogen fixation, process optimization for fermentor-based technologies, improved shelf life, better quality standards, setting up accredited quality control laboratories and standardization of GMP guidelines. Integrated nutrient management system would be further strengthened.

BIORESOURCES

The combined annual global market for the products derived from bioresources is roughly between US$ 500 billion and US$ 800 billion. India is one of the 12 global mega biodiversity centres harbouring approximately 8% of the global biodiversity existing in only 2.4% of the land area. The country is also home to two of the world's 25 hotspots. The varied cultural diversity across the country as well as a very ancient traditional knowledge system associated with the biodiversity represents added assets. Nonetheless, much of this biodiversity is in peril owing, in the main, to anthropogenic causes. Thus, if the goal of converting our bioresources-animal, plant, microbial and marine-into commercially useful products and processes is to be realized, we need to not only conserve the biodiversity and but also utilize it in a sustainable manner.

In this context, absence of a good quantitative information network on bioresources combining remote-sensing data and ground surveys is a major constraint. The situation is even worse for microorganisms. Field-and marine biologists rarely work with molecular scientists and chemists, pharmacologists or other experts, and there is practically no bioprospecting industry. While our traditional knowledge base would be the starting for bioprospecting, ethics and equity should be our guiding principles in benefit sharing.

ANIMAL RESOURCES

India is home to an estimated 86,874 species of animals accounting for 7.25% of global animal diversity. The degree of endemism is high and populations of several animal groups are diminishing due to habitat destruction and poaching. Several species, their products and the services rendered by them are crucial to our economic well-being: pollination services by insects (e.g., honey bees, bumble bees, moths, butterflies, beetles, flies) to our agricultural and forestry crops, honey, silk, lac, musk, skins are just a few examples.

Other species (e.g., molluscs, frogs, toads, spiders, termites, and snakes) represent potential reservoirs of useful products such as toxins, venoms, enzymes, therapeutic molecules and other bioactive substances. Prospecting for these and other products should be a priority. Biotechnology should be effectively employed for molecular characterization along with bioscreens in search of useful products. Utilization of selected species as bioreactors for production of complex proteins is another important opportunity.

PLANT RESOURCES

India has a huge treasure of plant resources with over 45,000 known species representing 11% of earth's flora. In terms of flowering plant diversity alone, India ranks tenth in the world. About 33% of flowering plants and 29% of total plants are endemic to the country. Genetic erosion is rampant and conservation is a priority. Prospecting of wild plant resources using molecular approaches and mechanism-based screening should be used to identify novel genes (temperature, drought, salinity tolerant) and gene products (therapeutic compounds, dyes, essential oils, biocontrol agents, gums resins and taxmins). There are potential ornamentals, including foliage-and flower-bearing plants that could be bulked up to be subsequently cultivated on large scale for domestic and international trade. Bioconversion-both cellular and microbial-should be employed to convert intermediates of secondary metabolism into valued added products. Application of genomics, proteomics and metabolomics in carefully selected plants will be very useful.

Biotechnology can contribute substantially in providing cost-effective therapeutically active biomolecules through target/mechanism-based screens, biotransformation, metabolic engineering and transgenic approaches. Biotechnology should also be utilized to add value to our traditional knowledge especially Ayurveda, Sidha and Unani systems as well as tribal and folk medicine. Medicinal plants are also the prime targets of bioprospecting. Besides, the tools of biotechnology can be used for conservation and characterization of plants.

Fossil fuels are chief contributors to urban air pollution and a major source of greenhouse gases (GHGs)-considered to be the main causes behind the climate change phenomena. In contrast, biofuels are renewable; hence, they can supplement hydrocarbon fuels, assist in their conservation, as well as mitigate their adverse effects on the climate.

Two major biofuels for the transport sector, bioethanol and biodiesel, are fast becoming popular in many countries around the world. While bio-ethanol (called ethanol) is produced from raw materials such as molasses, beet, sugarcane juice, grains and tubers, biodiesel is produced from oil (derived from oil-bearing seeds such as Jatropha curcas, Pongamia pinnata i.e.karanja).

India imports nearly 70% of its annual crude petroleum requirement. The net oil import bill (import minus exports) was Rs 77,058 crore (Rs 770.58 billion) in 2003-04 as against Rs 74,174 crore (Rs 741.74 billion) the previous year. This expenditure on crude purchase impacts the country's foreign exchange reserves in a big way. The petroleum industry now looks very committed to the use of ethanol as fuel.

It is estimated that 75% of the increase in world demand for oil will come from transport. India's transport sector will consume ever-higher amounts of fuel over the coming years. Being one of the largest producers of agro products, including sugarcane, India should take a lead in this worldwide effort at promoting sustainable development.

MICROBIAL RESOURCES

Currently only five percent microbes are culturable but there are others of considerable potential value that need to be characterized by new and novel techniques. The five percent culturable microbes have been a source of valuable products.

India should play a leading role in the study and utilization of microbial resources. Our priorities include: preparation of inventories based on primary and secondary data; exploration of micro flora in the north-eastern region of the country, and extreme habitats (hydrothermal vents, deep sea sediments, highly acidic, alkaline and anaerobic regions, degraded ecosystems etc.) for discovery of novel bioactive molecules; and study, characterization and screening of uncultivable microbes through appropriate molecular approaches.

MARINE RESOURCES

The economic zone of the sea as a source of novel genes and gene products-biopolymers, novel enzymes, new therapeutic leads, and other value-added products such as osmo-tolerant crops-has hardly been explored.

Marine organisms also present immense potential as biosensors for pollution monitoring as well as bioreactors for production of novel products. Besides, the study of deep-sea organisms including marine microbes has tremendous implications for human health. Expertise in these diverse areas is scattered across a number of agencies/institutions. Strategic Actions would be in the following areas.

Strategic Actions:

- There is an acute shortage of expertise in India particularly in taxonomy (the science of the classification of the living and extinct organisms) and microbial ecology. We need to take urgent steps to rectify this.

- Support to capacity building in microbial taxonomy through intensive training programmes at graduate and post-graduate levels
- Promotion of horizontal networking between remote sensing experts, field biologists and computer specialists for inventorisation of bioresources based both on primary and secondary sources of information
- Promotion of closer and effective interaction between biotechnologists, foresters, oceanographers and field biologists.
- Ensure that the use of bioresources be sustainable by regulating the harvesting of medicinal plants
- Formulate a policy to regulate the procurement and sale of medicinal plants in India. Introduce regulatory norms prescribed by DCGI that evaluate the efficacy, safety, and quality of herbal products, which currently are exempt from the scope of any regulation of the DCGI.
- Establish a close working relationship between field scientists, pharmacologists and clinicians so that an all round integration is achieved.
- Public-private partnerships need be promoted for product generation
- Creation of a gene bank for maintaining 'mined' genes
- There is, as on date, only one international depository authority (IDA) in the country at the Microbial Type Culture Collection (MTCC) at IMTECH, Chandigarh; however, for securing our IPR interests, we need to initiate steps to establish a few more centres as IDAs.
- Currently, MTCC does not accept biological materials such as cell lines, cyanobacteria, viruses etc. as it has no expertise or facilities for this purpose. Yet, these are essential for filing patents. IDAs in other countries may refuse to accept such material as they may be potentially hazardous or the shipments may have restrictions. In view of this, the scope of MTCC needs to be expanded by upgrading the existing expertise and infrastructure. Alternately, IDAs should be set up where such expertise and infrastructure are available.
- End products from bioprospecting need to be tested for a variety of parameters before commercial production can begin. There is a need to set up appropriate facilities for such late stage testing of products.
- An autonomous Centre for Marine Biotechnology is proposed to be set up under the auspices of DBT
- An autonomous Institute for Biotechnology for Herbal Medicine under the auspices of DBT is proposed to be established.

ENVIRONMENT

Environmental issues concern everyone. Biotechnology has tremendous potential for application to a wide variety of environmental issues including conservation and characterization of rare or endangered taxa, afforestation and reforestation. It can help in rapid monitoring of environmental pollution, eco-restoration of degraded sites such as mining spoil dumps, treatment of effluents discharged by industries (oil refineries, dyeing and textile units, paper and pulp mills, tanneries, pesticide units etc.), treatment of solid waste, and so on. A number of technologies have already been generated and demonstrated in the country. The real challenge is their adoption by the industry, which has been somewhat uneven.

In general, corporate groups have not been overly enthusiastic in adopting biotechnologies even where they have proven efficacy. The reasons may be several: industry is usually not involved at the planning stage of experiment; enforcement of environmental laws is not always strict or uniform at the ground level and offenders can often escape with impunity; manufacturers frequently change their production schedules based on demand profiles resulting in varied streams of effluents, but microbial consortia specifically designed to one set of effluents may be ineffective in breaking down the changed pollutants. The goal of environmental biotechnology would be to provide cost-effective and clean alternatives for risk assessment and quality monitoring, eco-restoration of degraded habitats, conversion of toxic recalcitrant chemicals into harmless by-products, bioremediation of wastes, value-added products from biomass, control of biological invasion through biotechnological interventions, greener process technologies, and effective ex situ conservation strategies. These can be fulfilled through a deeper understanding-and engineering-of the metabolic pathways for degradation of toxicants, environmental genomics and proteomics, and other molecular techniques.

Strategic Actions: For the diffusion of biotechnologies to be successful the following measures should be put in place:

- Ensuring effective and closer horizontal linkages between research workers and the user corporate groups
- Public-private partnership in research and application of clean technologies
- Strict enforcement of the 'polluter pays' principle. This would require interaction with law enforcement agencies
- Capacity building and training, through workshops, of law enforcement officials, municipal workers, state government functionaries and

corporate groups on role and relevance of biotechnology in waste treatment

- Steps to encourage small and medium business companies in producing eco-friendly products, microbial consortia etc. for wider usage
- Building greater awareness for protection of proprietary rights of microbial consortia through appropriate methods (e.g., process patent, trade mark etc.)

Greater inter-agency coordination between DBT, MoEF, ICAR, CSIR, CPCB, user agencies and industry through an inter-ministerial Task Force.

ALGERIA: AGRICULTURE AND INDUSTRY

The objective of providing full employment for the population has profoundly affected the basic options in the development of the Algerian economy. It led to stress on the rapid expansion of industry in order to create conditions for modernizing agriculture. By creating jobs in the secondary and tertiary sectors it was hoped to gradually reduce unemployment and under-employment and to have a significant impact on the size of the agricultural population in order to reduce pressure on the land. The stress on industry, in an initial phase of development, gave ample scope to the building-up of the sectors producing the production goods necessary for the modernization of agriculture, principally machinery, chemicals and petrochemicals. In a second phase, agriculture would become more integrated into the economy, and increase its capacity to develop its purchases *from* industry and its deliveries of agricultural raw materials to processing industries. Such a pattern of autocentred development is built on the hypothesis of a growth in peasant incomes fuelled by the growth of agricultural productivity, which was necessary 1) to raise the standard of living in the rural areas, and 2) to finance agriculture's demand from industry. To carry out such a process of growth and intensify relations between agriculture and industry requires three major conditions: That the global growth of agricultural production must be more than proportional to that of the numbers employed in agriculture. In other words, there must be net growth per person employed, and this is the source of improvements in incomes, ensuring a rise in both the standard of living of agricultural workers and agriculture's demand for industrial goods.

It must be possible to carry through the modernization of agriculture in conditions of productivity in the employment of the factors that ensure a minimum of profitability to capital invested. In other words, agriculture must be able to pay for what it buys from industry, not artificially, through continually rising agricultural prices, or subsidies from the state, but through the advances it achieves in productivity. Finally, what is produced must be at price levels sufficient to meet the needs of extended reproduction, but also compatible with consumers' purchasing power. If the opposite happens, the state is forced to act massively to keep consumer prices up at the expense of investment.

To make agriculture meet these three conditions may require phasing. Distortions may appear that must be corrected to enable agriculture gradually to adjust its relations with the rest of the economy but, in due course, agriculture must satisfy these key variables of the dynamic of its relations with industry.

The first variable is largely dependent on the pace of job creation outside agriculture, the capacity of industry and services to absorb the rural labour surplus. The quantities produced may have no effect on labour productivity if they are accompanied by a proportionate increase in the population employed in agriculture. In Algeria, the increase in output per worker cannot be obtained by increasing the area cultivated per unit of labour, as was partly the case in Europe, but must be the fruit of an improvement in the physical yields of crops and livestock farming. In that way the profitability of investment in production factors will be assured and hence that of backward-linked industrial sectors. In fact, as analysis of the Algerian case shows, in an autocentred development model, the growth of yields is the key to the whole dynamic of relations between agriculture and industry.

In the absence of advances in agricultural productivity, the rise in agricultural prices consequent upon the global stagnation of agricultural production may still fuel agriculture's demand for industrial goods, but at the price of a fall in real wages and the standard of living of the great mass of urban workers. To avoid this spiral, the state has to import food products that are in short supply on the market and subsidize producer prices for commodities in order to protect agriculture from external competition. The evolution of agricultural productivity rests essentially on agronomic and technical advances, the improvement of animal and plant species, and the introduction of new production methods. It requires that the state not only has the capacity to develop agronomic research and agricultural education and training programmes successfully but also be able to define and establish the social forms of agricultural modernization: in short, to define a *path to the modernization of agriculture* that meets the conditions of development.

THE ALGERIAN ECONOMY: DEVELOPMENT CHOICES

The fact that there are too many people engaged in agriculture and that it has little natural potential led to the emphasis in development choices on an initial phase of intensive industrialization. The priority thus given to industry was based on the hypothesis that no significant progress was possible in agriculture without an industrial policy to provide the motive force of the whole development dynamic. Industry was both to provide agriculture with the material means for its modernization (capital goods and intermediate products), and, over a long period, absorb the net increase in the agricultural population.

The first development plans of 1967, 1969 and 1970-73 aimed at keeping, 'for as long as possible', the numbers employed in agriculture at their 1966 level of about one million workers (1966 general population census). It was estimated that the total number of workdays supplied by agriculture was equal to some 700,000 full-time jobs only, meaning an underemployment rate among the agricultural population of about 30%. The analyses concluded that in order to improve productivity and incomes it was necessary to increase the number of workdays per worker.

LOW AGRICULTURAL POTENTIAL

With 7.5 million hectares of usable agricultural land, Algeria is poorly endowed with arable land, 0.3 hectare per inhabitant, and, given the high rate of population increase from now to the end of the current decade, the area available will not exceed 0.25 hectare per person. Much less than for Egypt in 1980, if account is taken of the double cropping in that country and biannual cropping in Algeria where cereal-fallow rotation occupies 80% of cultivated land. In fact 4.5 million hectares only are cultivated each year by one million workers, the remainder being rested. In assessing the productive agricultural potential, account must also be taken of the importance of natural constraints on production and the low productivity of the soils.

It is estimated that land in the plain or on slight or average slopes (less than 12.5%) with sufficient annual average rainfall, (higher than 500 mm per annum) to obtain relatively regular and satisfactory yields, occupies barely 2.7 million hectares or slightly more than one-third (36%) of cultivable areas and less than 10% of northern Algeria. Furthermore, even on this land, rainfall is both irregular from year to year and unevenly distributed over the year, control of water being generally essential to intensifying crops and ensuring regular production.

As always with high population density, cropping limits are largely exceeded if account is taken of the need for good soil conservation. In numerous regions, this density is excessive and leads to heavy soil degradation. As a result of the climate and slopes, erosion, both wind-and rain-induced, is considerable and the action of man on the environment simply accentuates a process which has doubtless been developing for centuries. Soil and productivity losses due to erosion are considerable and numerous cultivated areas ought urgently to be replanted with trees. An FAO study estimates that 30% of agricultural areas need protection works. It is estimated elsewhere that three million hectares need reforesting and a total of five million hectares need tree-planting in the north, as against a current rate of less than 10%.

Woodland, the other aspect of agricultural natural resources, is also seriously threatened with destruction. It covers only three million hectares of which only 2.3 are subject to forestry regulations. The reforestation programmes have so far been just sufficient to replace the areas destroyed by forest fires each year, but very much less than would be necessary to undertake to protect the most threatened areas effectively.

The quantity of sediments discharged into the sea each year amounts to 120 million metric tons, or the equivalent of several tens of thousands of hectares (50-60,000). Hydraulic features considerably increase the costs of bringing surface water into use, and reduce the life of dams.

In semi-arid regions where rainfall is regularly less than 500 mm, the inadequacy and irregularity of rainfall strongly influence yields, and conditions for a better economic use of land are closely dependent on irrigation. In Algeria, while potential water resources may appear relatively high, exploiting them proves to be limited and costly, as a result of the difficulties of the terrain and the narrowness of the coastal plains. The bulk of the runoff in the northern parts of the country, 12 billion m^3/year, simply flows into the Mediterranean. The drainage basins of the highlands (100,000 km^2) and the Saharan regions (100,000 km^2) with 0.7 billion m^3 each, run off only a little more than 10% of total rainfall. To surface water must be added exploitable subsoil resources, 1.7 billion m^3/year in the north, 2 billion m^3 in the Saharan regions. But these are only rather rough estimates, no complete inventory having so far been carried out and these figures probably do not reflect the real level of mobilizable resources and possibilities of exploitation, if one thinks of the reduction of available water stocks, the fall in ground water tables, the depletion of wells, and so on, observed in recent years. Exploitation of this potential is uneven, depending on the region and the nature of the water resources.

In the northern areas, only 5% of surface water is regulated with 0.6 billion m^3/year exploited, and 80% of ground water, with 1.3 billion m^3 exploited. In the Sahara. 30% of ground water is mobilized. 0.7 billion m^3 (50,000 hectares irrigated). There is also small-scale irrigation, surface water and off-takes from rivers, amounting to 0.9 m^3, but this figure seems too high. In all, only some 3.5 billion m^3/year is available, a very inadequate volume compared to other countries in the region. For example, Morocco has 10 billion m^3/year (800,000 hectares irrigated) and Saudi Arabia some six or seven billion m^3/year. Egypt 55 billion m^3. Agriculture receives about three-quarters of the available water, 2.6 billion m^3 for 260.000 hectares irrigated (95% in Morocco). The areas of the large, irrigated perimeters have remained unchanged for the last two decades, and the increase is essentially due to cheap small-scale irrigation projects: surface wells, hillside lakes, drawing water in the Oueds which further throws doubt on the figure for irrigated areas used in the statistics.

Drinking-water supplies (one household in two has running water) is the second biggest user with 760 billion m^3. Industry consumes only 4% of the total water available, far less than agricultural and human uses. The additional irrigation of 300,000 hectares planned by the year 2000 would require the mobilization of a further three billion m^3 or the building of 60 new dams and the digging of three million linear metres of wells-the cost would thus be very high.

Agricultural natural resources are, however, not limited to cultivable land. Algeria has vast areas of land, some 15 to 20 million hectares, situated between the 100 and 300 mm *isohyets,* devoted to livestock, essentially goats, which should be included in the country's production potential. It is of course true that degradation of pastures has by now gone very far and the average number of cattle per hectare cannot be more than one goat for five hectares (in some regions it is much lower than that-one goat for 10 hectares) with the result that the total number of goats that can be fed from the steppe's own resources is no more than four or five million head, with a net meat production in the region of 25 to 30,000 metric tons, indicating a very low productivity of soils and flocks. Environmental degradation is the consequence of replacing the old social forms of rangeland by ranching that facilitates rapid, intensive exploitation of resources for the market, which has led to the gradual impoverishment of small herdsmen and their eviction from the pastoral areas.

The evolution towards higher meat prices has led to a heavy concentration of the flocks in the hands of dealers who control the sheep market, and the decline of direct forms of exploiting flocks. The reduction of natural forage resources is leading to the clearing of the best soil for speculative barley-growing intended for feeding flocks. Yet despite the importance of the agro-ecological constraints, scope for agricultural progress does exist. The cropping system has remained extensive, mobilizing only limited land and labour resources, for yields per hectare that are excessively low. It is estimated that by bringing fallow land into production some two million hectares, 40% of the area cultivated each year, could be cropped. In addition, mobilization of water resources could significantly increase irrigated areas, thereby creating conditions for more intensive cropping. In the steppe, experiments have shown the possibilities of increasing forage production. Quite striking results have been achieved by setting aside areas closed to grazing animals covering 10,000 or 20.000 hectares. Forage production has thus been doubled or tripled and the number of livestock per hectare has moved in the same proportions; extending these results requires settlement of the problem of the customary use of rangelands prior to the introduction of techniques of regenerating vegetation and maintaining grazing areas.

Such a level of exploiting agricultural resources implies a heavy recourse to industrially produced factors of production to update equipment and cultivation methods.

CHARACTER AND SCALE OF INDUSTRIALIZATION

In the autocentred model of development, agriculture constitutes both a market for industrially produced capital and intermediate goods and a supplier of products *for* the foodstuff sector. Industrial development is thus governed by an internal dynamic and is not simply a means of maintaining the external balance or a supplement to the export economy (as in the cases of Tunisia and Morocco).

Algerian exports of manufactured or intermediate goods are in fact virtually non-existent-less than 2% of exports, comprising almost entirely hydrocarbons, which account for 98% of external receipts.

Industrialization programmes have given ample scope to sectors capable of supplying agriculture with the goods necessary to modernize its techniques and improve its productivity. As Table 1 shows, priority has been given to the agricultural equipment industries since the earliest phases of industrialization: A look at the dates when industrial projects were launched shows that most of those concerned with agricultural demand were initiated during the first phase of industrialization, 1966-71: agricultural machinery, fertilizers, irrigation equipment and so on.

Agricultural demand concerns a wide variety of branches of industry: upstream, steelworks and metal processing industries, metallurgical industries and machine industries (tractors, valves, pumps, transport equipment' agricultural implements), electrical industries (motors, cables for electrification) and finally chemical industries (fertilizers, chemical products, plastic items). Throughout the 1970s, the establishment of a wide range of industries (including hydrocarbons) took a large share of total investment: 55% for each period 1967-69 and 1970-73; 58% for 1974-77: 56.1%, 1978-79 and 32.6%, 1980-84,

Agriculture's shares were: 16%, 1967-69; 14.8%, 1970-73; 11%, 197477; 8.55%, 1978-79; and 5.2%. 1980-84.

Throughout the period, industrial investment narrowly defined amounted to approximately 28% of investments, the remainder (28-30%), more or less the same proportion, was absorbed by the hydro-carbon sector. It was thus a relatively modest percentage' compared to those of partly industrialized countries today (Yugoslavia. Mexico' Brazil).

Agricultural investment was around 10% of the total throughout the period, but a considerable proportion of industrial investment in the first and second phases of industrialization was devoted to the production of goods intended for agriculture. It should also be pointed out that the fall in industrial investment in the 1980s has not necessarily meant a shift to agriculture in relative terms; agriculture's share fell back by two or three points during the period 1978-84 compared to 1967-77.

The number of large production complexes set up to produce equipment and intermediate goods for agriculture is evidence of the stress put on the priority satisfaction of the needs of the agricultural sector; the tractor factory at Constantine (8.000 tractors a year), the agricultural machinery factory at Bel Abbes, the pump and valve factory at Medea, the diesel motor factory at Tizi Ouzou, the water pipe factory at Reghaia, and so on. In petrochemicals in particular, two large chemical fertilizer complexes were completed: one at Arzew (nitrogenous) and one at Annaba (phosphate), the operation of which required the building of several independent plants to manufacture intermediate products.

The Annaba complex includes a plant producing sulphuric acid with a capacity of 1.500 metric tons per day and one producing 500 metric tons per day of phosphoric acid. Two other plants producing multi-nutrient fertilizers at a rate of 500.000 metric tons per annum are in production in the same region and proposals to expand them during the period 1981-84 led to the creation of two new plants for nitric acid (400 metric tons per day) and ammonium nitrate (500 metric tons per day).

In plastics, the completion of a methanol complex in 1971 and a synthetic resin plant at Arzew, and a plastics complex at Skikda made it possible to create units producing sacks and sheets for agriculture. Production capacity is currently 100.000 metric tons per annum, with consumption for crops under plastic and packaging for irrigation pipes absorbing 31% of total production.

After a decade end a half of industrialization, the greater pert of agricultural demand for equipment and intermediate goods is being met by local production. Finally, for chemical products, an average of 90% of demand is met: phosphate fertilizer, 286.000 metric tons, and ammonium nitrate 170.000 metric tons. Agriculture's productive consumption of industrially produced goods has increased much faster than these industries' production capacities. Contrary to widespread ideas about Algeria' demand has constantly outstripped projections. As F. Yachir points out' this is true whichever industrial sector one looks at. Recourse to imports is still necessary because local supply is insufficient.

Projections in the 1969 industrial programmes on the needs of agriculture up to 1980 were easily surpassed by the consumption levels between 1970 and 1985. Agriculture more than doubled its equipment and purchases of intermediate goods between 1976 and 1985, compared to the periods 1966-75 and 1953-62. The total number of tractors in the country increased from 23,484 units in 1962 to 25,122 in 1973 and 60,000 in 1984. Fertilizer use has risen on average by 9.5% p.a. since the beginning of the 1970s: from 100,000 metric tons in 1970 to 205.000 metric tons in 1983-84.[12] The use of minor irrigation equipment, in the motor-pumps groups, also rose fivefold during the 1976-85 period, absorbing all local production.

Domestic supply is inadequate not only in terms of quantity, but agricultural equipment is often unsuitable and there is little variety of supporting equipment. Fertilizers and pest control products too, are not quite what the crops need.

Investments in sectors producing agricultural goods have thus been insufficient, judging from the high level of additional imports and the potential demand that programmes to step up agricultural production will create. In fact, despite the marked increase in intermediate consumption, the quantity of industrial inputs used falls far short of the technical norms deemed *necessary.*

The use of industrial inputs, especially fertilizers and agricultural machinery, compared to those of countries with intensive agriculture in the North, or a country such as China, is still relatively low. In Algeria, there is one tractor (60 hp) for 182 hectares and one combine harvester for 500 hectares.

The Ministry of Agriculture estimates that the use of fertilizers is only 44% of the quantities recommended to secure a marked increase in yields. The supply of agricultural equipment has grown by 6% to 8% p.a., but shortfalls are estimated to be 43% for tractors' 38% for combine harvesters and 65% for ploughing equipment.

Those responsible for agricultural development emphasize the important role of chemical fertilizers, mechanization, selected seeds and irrigation in raising yields, but this introduction of modern means requires changes in cropping methods and action by engineers and skilled workers to promote the use of *new production factors*.

The objectives of the four-year plans were to spread the use of herbicides on cereals and pulses, develop soil disinfection and intensify treatment with insecticides and fungicides in fruit-tree growing, but use of these products has reached only 40% of the recommended amounts.

To meet the demand from agriculture thus requires increased production of agricultural machinery, particularly low-powered tractors for smallholdings in the mountainous areas, and the diversification in supply of equipment and intermediate goods to meet the needs of diverse agricultural conditions.

PREVENTIVE AND THERAPEUTIC MEDICAL BIOTECHNOLOGY

A healthy population is essential for economic development. Important contributors to the total disease burden are infections like HIV-AIDS, tuberculosis, malaria, respiratory infections and chronic diseases affecting the heart and blood vessels, neuro-psychiatric disorders, diabetes and cancer. It is important to synchronize the technology and products with the local needs of the health system and to facilitate technology diffusion into health practice.

Increasing knowledge about pathogen genomes and subtypes, host responses to infectious challenges, molecular determinants of virulence and protective immunity and novel understanding mechanisms underlying escaped immunity and ways to develop novel immunogens will guide development of vaccines against infectious diseases. Translational research and ability to rapidly evaluate multiple candidates in clinical trials can help accelerate the pace of vaccine development.

New directions in manufacturing and delivery are emerging. Major opportunities to control costs are the more efficient processes for manufacturing of new pharmaceuticals, more efficient systems for production of therapeutic proteins and biomaterials and development of drug delivery systems that release drugs at a target site. A shift from parenteral to oral or transcutaneous administration of drugs and vaccine holds the promise of simplifying delivery in health systems. Medical biotechnology offers a significant possibility for Indian industry to establish a strong pharmacy sector, a growing number of small and medium biotechnology companies, a large network of universities, research institutes, and medical schools and low cost of product evaluation.

The medical biotechnology sector annually contributes over 2/3rd of the biotechnology industry turnover. The Indian vaccine industry has highlighted

India's potential by emerging as an important source of low cost vaccine for the entire developing world. Further, economic opportunities through contract research and manufacturing through global partnerships are large if supported by enabling government policies and incentives. The policy goal is to accord high priority to basic and applied research, to strengthen capacity in pre-clinical and clinical product evaluation technologies relevant to all aspects of health and medical care-predictive, preventive, therapeutic and restorative will be supported. Innovation will be supported through new granting mechanisms to support interdisciplinary networks and public private partnerships.

Strategic Actions:

(i) Research emphasis

- Basic and applied research would be supported in molecular and cellular biology, genomics, proteomics, system biology, stem cell biology, RNA interference, host response and new platform technologies.
- Pathogenesis of major diseases and molecular mechanics of disease transmission would be investigated
- Product development will be focused on vaccines, diagnostics, new therapies based on cell and tissue replacement, therapeutic antibodies, herbal medicine, plant based medicine, nucleic acids, therapeutics, drug and vaccine delivery systems, new anti microbial agents
- Research to improve production and manufacturing process and local production of biological reagents for development of diagnostics will be supported.

(ii) Improvements in infrastructure and networks

- A centre for translational research will be established. This new institute will be interdisciplinary and will deal with technology policy for public health, molecular pathogenesis of disease, technology development, scale up, product evaluation and technology diffusion into programmes. Centre will be unique in having a pool of scientists, physicians, engineers, and public health persons working on public health grand challenges. This institute will work through public-private partnerships and be a training centre for product development, IPR and regulation.
- A mission mode programme will be initiated in biomaterial and medical device area as an integrated effort by the Department of Science & Technology and Department of Biotechnology. The goal is to promote R&D and industrial activity.

- Two centres of molecular medicine will be supported within medical school system closely interacting with basic science institutes.
- A virtual network of stem cell centres will be established, using a city cluster approach to network scientists and clinicians. Two core stem cell research centres will be established together with several network sub-clusters. An umbilical cord stem cell bank will be established.
- Mechanism based screening of herbal drugs known in traditional Indian systems would be carried out so as to get value added therapeutics products quickly
- An inter agency task force of ICMR, Department of Biotechnology, and DST will be established to suggest strategies for strengthening medical school based research. Capacity related to translational biology, clinical trials, molecular epidemiology and product development would be strengthened. Integrated MD-PhD programs will be supported.

(iii) Streamline guidelines and procedures for the approval of recombinant pharmaceutical products. Currently there are multiple regulators, multiple ministries, lack of coordination between these regulators, Overlapping and duplication of responsibilities of these regulators, lack of a linear progression in the approval process and committees working outside their area of expertise. The Mashelkar Committee (2004) has drawn up a new procedural framework for Biopharmaceuticals, which has streamlined the regulatory process:

- IBSC will monitor all development work (upto 20 Litres) and recommend to RCGM for Animal Toxicity Tests (ATT) & Scale up.
- RCGM will evaluate the recombinant technology & grant permission for scale up-R & D, review and approve for preclinical animal toxicity tests and evaluate ATT data & recommend to DCGI for Human Clinical Trial (HCT).
- DCGI will permit Human Clinical Trials, review Human Clinical Trial Data, grant permission for Manufacture and Marketing the product and inspect the facility where product is manufactured.
- GEAC will review the manufacturing process to ensure that the LMO (living modified organism) is "inactivated" during the process and send its recommendations to the Drugs Controller General of India within the specified time. The GEAC would confine its approval role to LMOs and Category 3 and 4 microorganisms.

AGRICULTURAL PRODUCTIVITY FOR SUSTAINABLE FOOD SECURITY

Agriculture is the principal source of food, livelihood and foreign exchange earnings in sub-Saharan Africa (SSA) (Badiane and Delgado, 1995). Production is a particularly important component of food security in SSA; commercial imports will contribute less than 10 percent of food supply over the next ten years (Rosen, 1997). As a result, agricultural productivity is critical to SSA's ability to meet food security and economic development objectives in the face of rapid population growth. SSA is projected to have the world's highest population growth rate over the next decade, at 2.5 percent per year, with nearly three-quarters of its workforce employed in agriculture (World Bank, 1998).

Yet evidence of agricultural performance in SSA is mixed at best. *Total factor productivity* in agriculture is estimated to have grown by an average of 1.3 percent annually between 1961 and 1991 for Africa as a whole. *Land productivity* in SSA agriculture rose by an average of 1.9 percent per year between 1980 and the mid-1990s, while increasing by 3.4 percent and 2.0 percent annually in South Asia and Latin America and the Caribbean, respectively (World Bank, 1998). Over the same period crop production in SSA grew by 2.7 percent per year, and food production grew by 2.4 percent per year.

By contrast, *labour productivity* fell by an average of one percent per year in SSA agriculture. It increased by 1.9 percent and 2.5 percent per year, respectively, in South Asia and Latin America and the Caribbean. Complicating the differences in these indicators of agricultural productivity at the regional level are differences in the level and rate of change in each indicator across subregions and countries within SSA.

Table : Agricultural Productivity Levels and Trends

	Land productivity		Labour productivity		Labour/ Land	Total factor productivity	
	Level ($/ha, 1993)	Growth rate (%/year, 1980-1993)	Level ($/worker, 1995)	Growth rate (%/year, 1980-1995)	Ratio (workers/ ha, 1991)	Level (index, 1991)	Growth rate (%/year, 1961-1991)
Sub-Saharan Africa	68	1.9	392	-1.0	0.30[a]	0.8[a]	1.3[a]
Central Africa							
Cameroon	313	1.7	827	-0.3	0.17	0.9	1.8
Central African Rep.	119	1.7	516	0.8	0.18	0.6	2.7
Congo	28	2.2	629	1.0	0.04	0.9	1.2

Gabon	74	0.8	1 516	0.5	0.05	0.3	-2.3
Zaire	113	2.4	219	0.0	0.34	0.6	8.1
East Africa							
Burundi	270	1.9	177	-1.4	1.01	2.9	3.4
Ethiopia	116	na	181	na	0.22	0.2	-1.7
Kenya	90	2.2	240	-0.7	0.13	0.6	1.9
Madagascar	34	2.1	178	-0.4	0.09	0.7	-0.1
Rwanda	378	-1.2	206	-2.6	1.42	0.7	6.1
Uganda	515	na	592	na	0.65	2.9	7.8
Sahel							
Burkina Faso	93	2.9	182	1.1	0.25	0.1	0.8
Chad	10	4.0	198	2.0	0.16	0.6	0.2
Gambia	199	1.6	167	-1.7	1.00	0.4	-1.5
Mali	33	2.5	259	0.2	0.06	0.8	0.8
Mauritania	7	2.6	na	na	0.01	0.1	-0.3
Niger	63	0.8	256	-0.9	0.20	0.7	1.5
Senegal	118	1.9	375	0.9	0.34	0.6	1.5
Somalia	na	na	na	na	0.05	0.8	1.2
Sudan	na	na	na	na	0.04	0.8	0.1
West Africa							
Benin	321	4.2	563	2.8	0.57	1.1	1.2
Côte d'Ivoire	212	0.6	1 354	-0.8	0.13	1.1	0.9
Ghana	227	0.4	684	-1.1	0.29	0.7	-0.5
Guinea	54	na	225	na	0.24	0.5	1.2
Guinea-Bissau	78	2.9	292	3.1	0.21	0.1	-2.1
Liberia		na	na	na	0.08	1.1	0.0
Nigeria	150	2.3	684	2.4	0.28	0.6	-0.3
Sierra Leone	123	0.4	344	-0.4	0.30	0.5	0.5
Togo	189	3.6	461	0.9	0.32	0.4	-1.3
Southern Africa							
Angola	9	na	149	na	0.07	0.4	-0.8
Botswana	5	1.7	483	1.4	0.01	0.2	1.3
Lesotho	24	-2.9	194	-2.7	0.23	0.1	-1.7
Malawi	153	0.4	156	-0.3	0.69	0.7	0.3
Mozambique	12	na	92	na	0.11	0.3	0.3
Namibia	9	0.9	1 458	0.8	0.004	1.4	1.0
South Africa	49	0.7	2 870	1.3	0.02	1.4	1.3

Swaziland	na	na	na	na	0.12	1.2	3.3
Tanzania	na	na	na	na	0.20	0.4	0.2
Zambia	7	1.2	100	-1.0	0.04	0.5	1.5
Zimbabwe	41	1.5	266	-0.7	0.27	0.7	2.0

Sources: Lusigi and Thirtle (1997), World Bank (1998). [a]Averages for all Africa. "na" no data available.

TRENDS IN AGRICULTURAL PRODUCTIVITY IN SUB-SAHARAN AFRICA

Table 1 presents three measures of agricultural productivity and their rates of change, by region and country, based on data from the World Bank (1998) and Lusigi and Thirtle (1997). It is important to note that the value of agricultural output used in generating these statistics is itself difficult to estimate because a large share of agricultural output is either consumed within the household that produces it, or is exchanged for commodities other than money (World Bank, 1998). The value of output is often estimated indirectly using a combination of methods, including reliance on estimates of yields and area under cultivation. To the extent that such attempts underestimate the true magnitude of agricultural output, estimates of agricultural productivity will themselves be biased downwards.

Output per unit of land, or crop yield, is commonly used by agricultural scientists to assess the success of new production practices. Land productivity is also used by national policy-makers to assess agricultural production for meeting national food security needs. Output per agricultural worker, on the other hand, may be a more important indicator of rural standards of living and welfare (Block, 1995). Recognizing that food may be acquired through exchange as well as production, income becomes an important determinant of access to food and thus of food security. As a result, labour productivity may be particularly important as an indication of the ability of agricultural workers to acquire sufficient food, whether or not they produce food themselves.

Land productivity averaged US$68 per hectare of agricultural land (measured as the sum of arable land, permanent cropland and permanent pasture) for SSA as a whole in 1993, compared with $519 in South Asia and $116 in Latin America and the Caribbean. Values ranged from $5-10 per hectare in the drier countries of Southern Africa and the Sahel to $200 per hectare and more in the East African highland countries and tropical West Africa. For SSA as a whole, land productivity grew at an average rate of 1.9 percent per year between 1980 and 1993, with slow to moderate growth in most countries. Land productivity grew most rapidly in the Sahelian countries and West Africa, and more slowly in Eastern and Southern Africa (Table 1). These patterns in land productivity levels and growth are also depicted in the second map in Figure 1.

Labour productivity averaged $392 per agricultural worker for SSA as a whole in 1995, this compares with $383 in South Asia and $2 292 in Latin America and the Caribbean. Values ranged from $100-$200 per worker in many countries in Eastern and Southern Africa and the Sahel to more than $500 per worker in parts of West and Central Africa. Labour productivity declined at an average rate of 1.0 percent per year for SSA as a whole between 1980 and 1995, with modest growth in West Africa and declines in Eastern and Southern Africa (Table 1). The first map in Figure 1 illustrates patterns of labour productivity. In contrast to the pattern evident for land productivity, the most rapid growth in labour productivity appears to be taking place in those countries that already have the highest levels of labour productivity. Perhaps this reflects the pull of off-farm employment opportunities, suggesting that disparities in labour productivity across countries may increase over time.

Low (or declining) labour productivity is consistent with high (or growing) land productivity in the context of a large (or expanding) agricultural labour force. Such patterns are evident in the agricultural labour/land ratios reported by Lusigi and Thirtle (1997) and presented in Table 1. The labour/land ratio is generally high in East Africa and low in Central and Southern Africa and the Sahel.

Land and labour productivity measures are both incomplete indicators of agricultural productivity, since they measure the productivity of only a single factor of production, and may well move in opposite directions. In an effort to address this problem, economists estimate total factor productivity (TFP), which measures changes in agricultural output relative to changes in an aggregated index of multiple inputs. TFP growth reflects factors such as technical change or improvements in infrastructure or research. It can also reflect failure to include or measure inputs such as depletion of soil or other natural resources. The last two columns of Table 1 report estimates by Lusigi and Thirtle (1997) of TFP levels for 1991 and rates of change for 1961-91. For Africa as a whole the TFP index estimated by Lusigi and Thirtle averaged 0.8. This indicates that on average, the productivity of the set of inputs measured was 0.8 times as high in 1991 as it was in the most efficient countries in 1961, up from an average of 0.7 in 1961. Levels are generally mixed, with the highest estimates in Uganda and Burundi, suggesting the importance of land quality, availability of water, and labour supply in driving agricultural TFP (Lusigi and Thirtle, 1997). Uganda also has one of the highest rates of growth in TFP, averaging 7.8 percent per year since 1961. For Africa as a whole, TFP grew at an average rate of 1.3 percent per year over the period.

DATA AND MEASUREMENT ISSUES IN ESTIMATING SSA AGRICULTURAL PRODUCTIVITY

Data on agricultural outputs and inputs are costly to collect. Sub-Saharan African countries have limited budgets devoted to data collection, with the result that data on both conventional and non-conventional inputs are often unavailable or incomplete. For example, in the United States, the definition of conventional inputs has expanded to include pesticides, energy, feed, seeds and intermediate livestock inputs. When an input such as pesticides is left out, increased output that might be attributed to increased pesticide use may be incorrectly attributed to TFP growth instead.

Inadequacies in the international data set for productivity analysis have been pointed out by a number of researchers. To date, most research has concentrated on measuring productivity in new ways with the same existing and insufficient data. Efforts have been made in recent years to improve the data, such as constructing a data set of public agricultural research expenditures by country, but much work still needs to be done.

Aggregating Agricultural Output

In order to aggregate agricultural output for international consistency, output must be measured in a common unit. Typically, output has either been reported in terms of dollars or in terms of "wheat units". The wheat units approach was developed by Hayami and Ruttan (1985) and is based on the ratio of each individual commodity price to the price of wheat in India, the United States and Japan. Official exchange rates are generally considered to be a poor choice for converting output in local currency units to dollars due to the biases introduced by fixed exchange rates or sudden devaluations. Most researchers use the purchasing power parity exchange rates inherent in the Food and Agriculture Organization of the United Nations (FAO) international dollar concept. However, Block (1995) argues that wheat units are preferable due to the impact of annual price movements that can affect the agricultural value added to which the international dollar conversions are applied. However, most recent studies have used FAO's international dollars.

CONVENTIONAL FACTORS OF PRODUCTION

Conventional inputs to agricultural production are land, labour, physical capital, livestock and fertilizer. For international comparison studies, the source of most data on these inputs is FAO. Conventional inputs are typically measured in relatively simple physical terms that mask potentially important qualitative variations.

Land is typically measured as hectares of agricultural land, i.e. arable and permanent cropland and permanent pasture. FAO statistics indicate that Africa had just over one thousand million hectares of agricultural land in 1990, up 0.1 percent annually in the previous two decades (Table 2). This measure does not account for land quality. Failure to account for land quality may lead researchers to incorrectly attribute to other inputs differences in production that are actually due to differences or changes in land quality. Some attempts have been made to control for differences in land quality by including a land quality index as a non-conventional input. Such attempts are noteworthy (Craig, Pardey and Roseboom, 1997), but they have been able to apply only one land quality indicator per country. This is problematic for large countries that span several ecozones.

Table : Levels and Trends in African Agricultural Output and Inputs, 1970-1990

	Level		Growth rate
	1970	**1990**	**(percent/year, 1970-1990)**
Agricultural output (*index, base 1989-1991*)	66	98	2.0
Agricultural land (*millions of hectares*)	1 059	1 090	0.1
Agricultural labour (*millions*)	120	167	1.7
Tractors in use (*thousands*)	334	521	2.2
Cattle (*millions*)	149	188	1.2
Fertilizer consumption (*thousands of tonnes*)	1 615	3 686	4.2

Source: FAO (1999).

The proxy for agricultural labour is often the economically active population in agriculture. Early FAO statistics only included males in the agricultural labour force. More recent data have included both males and females. Table 2 shows that Africa's agricultural labour force grew 1.7 percent annually between 1970 and 1990, to 167 million. However, this agricultural labour force variable still does not control for differences across countries in the composition (and thus potentially the quality) of the agricultural labour force by age and education. An additional problem with the FAO data is that the economically active population in agriculture is defined to include workers in agriculture, forestry and fisheries. This implies that the number of workers is overstated for every country, and is more heavily overstated for countries with large forestry and/or fishery sectors relative to their basic agriculture sectors. A few researchers have made attempts to correct for the quality of the agricultural labour force by including national-level measures of education or literacy as non-conventional inputs. No researchers of SSA have adjusted the quality of labour directly by sex, age and education.

Another problem is that many of these economically active agricultural workers are not employed full-time in agriculture. Evidence from Africa suggests that many farmers are heavily involved in off-farm work to supplement their farm incomes. In such cases, the agricultural labour force may look unduly large and thus bias estimates of labour productivity downwards.

The use of physical capital is typically measured by the stock of tractor horsepower (Hayami and Ruttan, 1985). The number of tractors in use in African agriculture increased 2.2 percent annually between 1970 and 1990, to 521 thousand (Table 2). Such a measure is problematic in SSA where many farmers continue to use hand implements, especially in hilly regions where tractors are ill-suited. Farmers' investments in hand hoes, carts, ploughs, fencing, buildings and other locally produced capital inputs have not been accounted for in national and international productivity studies. Incomplete measurement of physical capital, in terms of quality and quantity, will bias productivity estimates. In an effort to improve measurement of physical capital, Craig, Pardey and Roseboom (1997) have updated the physical capital variable to include two-and four-wheel tractors that are converted to horsepower using regional averages.

Livestock is a difficult input to measure since livestock may serve as both an input and an output in agricultural production. As an input, livestock has been measured as the number of livestock on farms at a given point in time. Kawagoe, Hayami and Ruttan (1985) included all livestock as an input, arguing that they represent long-term capital formation in the agricultural sector. Arnade (1997) did not include livestock as an input to production, arguing that in developing countries that do not have meat-processing sectors, livestock is usually sold directly as an output. Craig, Pardey and Roseboom (1997) included livestock as an input, but only included those animals that are primarily used for traction or breeding services. Clearly, differences in how the livestock variable is treated will affect estimates of the levels, sources, and changes in agricultural productivity. Table 2 shows a total of 188 million cattle in Africa in 1990, up 1.2 percent annually since 1970.

Commercial fertilizer inputs are measured as tons of nutrient units of nitrogen, phosphorus and potash. FAO data indicate that fertilizer consumed in African agriculture in 1990 totalled 3.7 million tonnes, up 4.2 percent annually since 1970 (Table 2). Fertilizer consumption subsequently declined by 1.1 percent per year between 1990 and 1995. However, the fertility benefits of organic sources of nutrients, such as manure and legumes, are not accounted for in this measure. This omission is potentially significant given widespread reliance on organic sources of nutrients in SSA.

Non-Conventional Factors

Non-conventional factors include private and public agricultural research, education, infrastructure, government programmes and policies, and environmental degradation. Sometimes in an attempt to adjust conventional inputs for quality, researchers have included these variables in the set of non-conventional inputs. Examples include land quality indicators (Frisvold and Ingram, 1995), or proxies for agricultural labour quality, such as literacy and life expectancy (Craig, Pardey and Roseboom, 1997).

At the national level, public agricultural research expenditures are generally used as a proxy for research and development. Public agricultural research expenditures are typically lagged for a number of years to compensate for the time required for research to reach fruition. However, this measure does not account for the spillover of research that is easily transferred from other countries. Private research expenditures have not been included in studies of developing countries since that information has not been collected.

Education is related to the quality of the agricultural labour force. For example, literacy would be expected to improve a farmer's ability to make use of information provided by extension services, or to keep better track of the costs and returns to alternative inputs or marketing opportunities. More generally, a more educated populace may also provide better services to agriculture, improving agriculture's productivity even without changing the quality of the agricultural labour force directly.

Since no data are available specifically on the educational level of the agricultural labour force in most countries, national-level proxies are used. Education has been measured by the school enrolment ratio or the adult literacy rate.

More generally, the overall quality of the labour force has been measured by national life expectancy (Craig, Pardey and Roseboom, 1997) and by historic calorie availability (Frisvold and Ingram, 1995). In an effort to focus more specifically on the education achievements of the agricultural labour force, Hayami and Ruttan (1985) also looked at the number of agricultural college graduates as a proxy for the level of advanced technical education in agriculture.

Public investments in infrastructure such as roads, utilities, and communications can increase agricultural productivity as well, by lowering the cost of inputs at the farm level and increasing farmers' access to marketing opportunities. Proxy variables include paved road density or by gross domestic product of each country's transportation and communication sectors (Hu and Antle, 1993).

Government programmes and policies also affect agricultural productivity. For example, Fulginiti and Perrin (1993) argue that historic agricultural output and input prices affect the technology chosen by farmers, and thus drive observed productivity patterns. Prices may be affected by government policies that tax or subsidize agriculture, and a "net protection coefficient" is used to capture the effect of these policies on agricultural productivity by Hu and Antle (1993) and Fulginiti and Perrin (1997). Block (1995) used depreciation of the real exchange rate as a proxy for government policy reform. The past export growth rate and export instability (Frisvold and Ingram 1995) have also been used as proxies for government policies that might affect productivity. They argue that export growth tends to stimulate overall economic development and productivity growth. They also note that export instability might slow productivity growth.

Researchers have used several variables in an attempt to adjust for the impact of land quality differences on productivity. Several studies have used a land quality index created by Peterson (1987) that indexes land quality at the national level as a function of historic precipitation and the share of a country's land area devoted to pasture and crops. Researchers have also used the percentage of a country's land that is arable, the percentage of land that is irrigated and mean rainfall to adjust for variations in land quality across countries.

Environmental degradation and actions that farmers take to reduce or reverse degradation have been recognized as potentially significant inputs to the production process (Thrupp, 1997), but they have not yet been measured and included as explanatory variables in productivity studies due to the scarcity of nationally or internationally comparable data.

EXISTING RESEARCH ON SUB-SAHARAN AGRICULTURAL PRODUCTIVITY

Research on cross-country agricultural productivity comparisons has concentrated on two areas. The first type of studies has estimated econometrically multi-country aggregate or meta-production functions to explain cross-country variation in agricultural productivity and to estimate production elasticities. The second, more recent type of studies has used data envelopment analysis (DEA) to construct Malmquist TFP indices to show which countries are experiencing the highest or lowest rates of growth.

Aggregate Production Functions

Table 3 compares estimated Cobb-Douglas coefficients from studies that estimated separate equations for developed countries (DCs), developing countries (LDCs) and African countries. Kawagoe, Hayami and Ruttan (1985) split their sample of 43 countries into 21 DCs and 22 LDCs. They found all the conventional variables as well as technical education to be important in explaining output levels for the DCs. For the LDCs, land and fertilizer were not found to be significant explanatory variables, but livestock was more important when compared to the DCs. In addition, both the school enrolment ratio and technical education were found to be significant. Summing over the coefficients of conventional inputs, Kawagoe, Hayami and Ruttan found increasing returns to scale in the DCs and constant returns to scale in the LDCs. They explain that an increasing substitution of large machines for labour has led to the pattern of increasing returns to scale observed in DC agriculture. On the other hand, they argue that many of the productivity increases realized in LDC agriculture have come from the increased use of high-yielding seed varieties and fertilizer, which are scale-neutral. They conclude that there is still much scope in LDC agriculture to increase productivity, and in particular labour productivity, by increasing investment in education, research, and modern inputs.

Fulginiti and Perrin (1993) conducted a study of 21 countries that included three African countries. They were particularly interested in examining how past pricing policies affected technology choice and thus current agricultural productivity. In addition to conventional inputs, they included past output prices, past wages and past fertilizer prices as well as the school enrolment ratio, a land quality index and research expenditures as non-conventional inputs. They found all of the conventional inputs to be significant as well as fairly close to the range of estimates found in other studies, except for land, which had a larger coefficient (0.25). They argued that previous studies, which did not include country-specific effects such as the land quality index used in their study, tended to bias the land elasticity downward. Higher school enrolment ratios are associated with higher productivity. The coefficient on the past output price variable was 0.13, indicating that a one percent change in past output price expectations would lead to a 0.13 percent shift in the production function. Echoing Hu and Antle's (1993) study of non-African countries, they argue that eliminating policy interventions that tax agriculture would increase productivity, while elimination of subsidies would decrease productivity.

Fulginiti and Perrin (1993) also estimated productivity increases that would be possible from eliminating output price policies that tax agriculture. The estimated productivity increases associated with such policy reforms ranged from 1.4 percent in Colombia to 129 percent in Zambia. Côte d'Ivoire and Ghana had estimated productivity increases of 88 percent and 16 percent, respectively. Fulginiti and Perrin tried several different formulations of the research expenditure variable, and found it to be significant only when measured as the work-years of government agricultural research, cumulated into a stock using a five-year lag. The sensitivity of these results with respect to different variable specifications reinforces the importance of measurement issues in evaluating agricultural productivity.

Frisvold and Ingram (1995) estimated land productivity for 28 countries in sub-Saharan Africa between 1973-1975 and 1983-1985. They estimated land productivity grew at an annual rate of 1.5 to 1.8 percent in most regions over the period. All conventional inputs were found to be significant, and the coefficient on labour was particularly large at 0.59. Frisvold and Ingram found that increased application of agricultural labour was the single most important factor in explaining growth in land productivity, and concluded that substantial increases in land productivity should not be expected until land becomes relatively scarce, echoing Binswanger and Pingali (1988) and Boserup (1965). They also found land quality, as measured by the Peterson (1987) index, to be an important explanatory variable, as was the percentage of area irrigated. Similar to Fulginiti and Perrin (1993), they did not find research expenditures to be significant. However, they did show that export growth and historic calorie availability have contributed positively to productivity growth in SSA. Growth in the stock of conventional inputs as a whole accounted for more than two thirds of growth in land productivity, which in turn accounted for the majority of growth in agricultural output.

Lusigi and Thirtle (1997) used a sample of 47 countries in Africa over the period 1961-1991. They estimated an average rate of TFP growth of 1.3 percent per year over that time period. Livestock, labour and land are the most important inputs explaining variation in productivity across the countries, with population pressure (and thus expansion of the agricultural labour force) an important factor driving productivity growth. Fertilizer and machinery are significant but of less importance. Like Frisvold and Ingram (1995), Lusigi and Thirtle stressed the contribution of population pressure to faster growth, arguing that land abundance depresses farmer incentives to increase land productivity by adopting yield-increasing technologies. They also found land quality and research expenditures to be important explanatory variables.

While the significance of specific inputs varies across studies using the production function approach, these studies have found in general that increased use of conventional inputs, especially those other than land, are the most important factors in explaining growth in agricultural productivity and output. Studies that have included measures of infrastructure and of land and labour quality have shown that variation in those factors is also important. The estimated effect of research on agricultural productivity varies by region studied, how the research variable is specified, and on the lag length selected.

The Malmquist Index

A traditional Tornqvist TFP index, which requires data on input prices and is used for calculating agricultural productivity in the United States, has never been constructed for SSA or used in multi-country production studies because the required data are not available. Instead, studies of agricultural productivity in SSA have used either a production function approach or a Malmquist TFP index, neither of which requires data on input prices. The Malmquist TFP studies are relatively new and have provided some insight into the relative ranking of countries in terms of productivity. However, researchers have not yet attempted to explain the differences in productivity across countries.

Data envelopment analysis (DEA) for creating Malmquist indexes of total factor productivity is a non-parametric programming approach that uses data on physical inputs and outputs to model efficiency levels. DEA identifies the best-practice countries out of a set of countries under study. The best-practice countries are those which minimize inputs per unit of output. Those countries then define the efficient production frontier. The efficiency of all other countries under study is measured relative to that efficient production frontier. Technical change is determined by the position of a country relative to the changing position of the efficient frontier over time.

Since the efficient production frontier depends on the set of countries included in the study, the productivity growth rates calculated by each study will depend on the countries included in the analysis. For example, Table 4 compares the TFP growth rates calculated from Malmquist indexes for selected SSA countries included in three different studies. Lusigi and Thirtle (1997) calculated the Malmquist TFP growth rates based on a set of 47 African countries for the period 1961-1991. Trueblood and Coggins (1997) constructed their

Malmquist TFP growth rates by analysing 117 countries from all world regions for 1962-1990. Thirtle, Hadley and Townsend (1995) estimated growth rates for 22 SSA countries over the period 1971-1986. The differences in results due to the selection of countries and time periods are striking. Relative to the efficient frontier for Africa alone, Lusigi and Thirtle (1997) show that the majority of SSA countries had positive TFP growth rates. Zaire, Uganda, Rwanda and Burundi (3.4 to 8.1 percent per year) achieved the highest rates of growth. In comparison, Trueblood and Coggins (1997) showed negative TFP growth rates for most SSA countries over the period 1962-1990 when they were compared to a global set of countries encompassing a much wider range of technologies, suggesting the technology gap between Africa and the rest of the world is widening. The ranking of SSA countries with positive growth rates were Congo (Brazzaville), South Africa, Uganda, Benin, Sierra Leone and the Central African Republic. Two of the highest-ranking countries in the study by Lusigi and Thirtle (1997), Rwanda and Burundi, had nearly the lowest growth rates of TFP in the worldwide study (-13.9 and-16.4 percent, respectively). Thirtle, Hadley and Townsend (1995) find low but positive TFP growth rates for most of the 22 SSA countries they studied for 1971-1986, with the highest rates in Rwanda and Burundi.

Table : Comparisons of Malmquist TFP Growth Rates for African Countries

	Lusigi & Thirtle	**Trueblood & Coggins**	**Thirtle, Hadley & Townsend**
	(percent/year, 1961-91)	**(percent/year, 1962-90)**	**(percent/year, 1971-86)**
Central Africa			
Cameroon	1.8	-2.8	1.0
Central African Republic	2.7	0.0	1.8
Congo	1.2	1.6	-1.4
Gabon	-2.3	-25.8	na
Zaire	8.1	-2.1	1.5
East Africa			
Burundi	3.4	-16.4	2.7
Ethiopia	-1.7	-2.5	0.7
Kenya	1.9	-0.2	0.6
Madagascar	-0.1	-5	na
Rwanda	6.1	-13.9	3.2

Uganda	7.8	1.2	na
Sahel			
Burkina Faso	0.8	-4.6	1.6
Chad	0.2	-2.5	na
Gambia	-1.5	na	na
Mali	0.8	-4.1	2.2
Mauritania	-0.3	-12.3	na
Niger	1.5	-5.7	na
Senegal	1.5	-3.4	-0.0
Somalia	1.2	-0.3	-0.3
Sudan	0.1	-2.5	-0.3
West Africa			
Benin	1.2	0.8	na
Côte d'Ivoire	0.9	-1.0	1.3
Ghana	-0.5	-2.5	0.6
Guinea	1.2	-1.7	na
Guinea-Bissau	-2.1	na	na
Liberia	0.0	-4	na
Nigeria	-0.3	-4.7	1.0
Sierra Leone	0.5	0.4	0.3
Togo	-1.3	-6.4	0.1
Southern Africa			
Angola	-0.8	-3.2	na
Botswana	1.3	-0.1	na
Lesotho	-1.7	-2.7	na
Malawi	0.3	-3.1	0.6
Mozambique	0.3	-1.0	na
Namibia	1.0	na	na
South Africa	1.3	1.5	na
Swaziland	3.3	na	na
Tanzania	0.2	-3.0	2.0
Zambia	1.5	-1.9	-1.0
Zimbabwe	2.0	-0.4	0.3

Sources: Lusigi & Thirtle (1997); Trueblood & Coggins (1997); Thirtle, Hadley & Townsend (1995).

Thirtle, Hadley and Townsend (1995) decompose the Malmquist TFP growth rates they find for 22 SSA countries into technical progress (from the time series for this panel of countries) and efficiency changes (from the cross-section). Investments in infrastructure, extension and the level of real protection on international agricultural markets are shown to be significant in explaining efficiency change, while tractors, the labour-land ratio, R&D and secondary education are found to explain the variation in technical progress.

They find the labour-land ratio, or population density, to be the single most important explanatory variable, again suggesting that productivity growth will accelerate in land abundant countries as population density increases.

GROWTH ACCOUNTING

Paralleling the two types of productivity studies, aggregate production function and TFP indexes, are two types of growth accounting exercises. In the first case, the coefficients from the aggregate Cobb-Douglas production function are used to calculate the percentage growth in output due to changes in the level of each input. Changes in the residual or TFP are attributed to non-conventional factors.

Arnade (1997) conducted this exercise for a set of 77 countries, including five in Africa (Kenya, South Africa, Sudan, Zaire and Zimbabwe). He included livestock as a conventional factor only in Asia, where bullocks are commonly used for ploughing. He did not include livestock as a factor of production in other regions, arguing that there is little meat processing in the developing countries so livestock is sold as an output.

For the Cobb-Douglas coefficients, Arnade used 0.20 for labour, 0.45 for land, 0.15 for fertilizer, and 0.20 for tractors. Constant coefficients such as these are consistent with marginal factor productivities that may vary with input levels across countries and across time. Arnade's coefficients are within the range of estimates found in the empirical studies reported in Table 3, except for the coefficient on land, which is larger than those found in other studies. This is likely due to the exclusion of livestock from the aggregate production function.

Results from Arnade's study for the five African countries and a representative sample of other countries are reported in Table. As expected, during the period 1961-1987, the growth rate of productivity was much higher in the "advanced technology" countries than in Africa or Asia. In the advanced technology countries, the level of land and labour inputs dropped while output increased, due mainly to increased fertilizer use and improvements in productivity. In the Asian countries reported, the contribution of land expansion to output growth was low, while tractor use showed a high rate of increase.

The growth rate of labour and fertilizer was moderate, while productivity growth was mixed. For the African countries, the increases in agricultural land were relatively small, while increases in labour, tractor, and fertilizer use contributed significantly to output growth.

The differences between Asia and Africa and the advanced technology countries are stark. The latter have been able to maintain increases in agricultural output while decreasing labour and land inputs. Asia and Africa have relied much more heavily on increases in the conventional inputs.

This is hardly surprising, given their relatively low levels of conventional input use, other than labour and land.

For Africa, excluding South Africa (because of its advanced agricultural technology relative to the rest of SSA) and Zambia (due to its negative growth rate of productivity), growth in the use of conventional inputs explains approximately 60 to 80 percent of the growth in agricultural output over the period under study.

However, Craig, Pardey and Roseboom (1997) noted that the conventional inputs in their equation accounted for 72 percent of the variation in productivity, both for LDCs alone and also when DCs were included. Frisvold and Ingram (1995) did perform an in-depth growth accounting study for their sample of 27 SSA countries.

For three of the four SSA regions under study (the Semi-Arid Tropics, the Sub-Humid Tropics and the Humid Tropics), they found that conventional inputs explained 84 to 93 percent of the increase in agricultural output. For the Lowland Humid Tropics (which consisted only of Mauritius and Madagascar) the figure was only 22 percent.

PROJECTIONS OF FUTURE GROWTH IN AFRICAN AGRICULTURAL PRODUCTION

A number of research efforts have generated projections of future growth in agricultural production in SSA. For example, FAO projects that gross agricultural production in SSA will increase at an average annual rate of three percent between 1988-1990 and 2010 (FAO, 1993). Of this increase, 53 percent is projected to come from increases in yield, 30 percent from increases in arable land, and 17 percent from increases in cropping intensity.

The International Food Policy Research Institute projects similarly strong growth in grain production in SSA, averaging three percent annually between 1990 and 2020. Area in cereals is projected to increase 1.3 percent annually over the period, and yields are projected to increase 1.7 percent per year.

World Bank simulations for selected developing countries include projections of 2.5 percent annual growth in grain production in Central Africa between 2000 and 2010, and 1.5 in Nigeria and South Africa, from a combination of yield and area increases (Mitchell and Ingco, 1995).

The USDA's Economic Research Service (Shapouri and Rosen, 1998) projects that food production in SSA will grow at an average rate of 2.3 percent per year between 1995-1997 and 2008 through a combination of area expansion (1.3 percent per year) and yield increases (1.0 percent per year).

Table : Contributions to Average Rates of Output Growth, Based on Coefficients from an Aggregate Cobb-Douglas Production Function, 1961-1987 (Growth Rates in Percent per Year)

Country	*Output*	*Labour*	*Land*	*Tractors*	*Fertilizer*	*Productivity*	*Animals*
C-D coefficients		0.20	0.45	0.20	0.15		
Africa							
Kenya	2.81	0.67	0.13	0.32	1.09	0.60	
South Africa	2.14	-0.19	-0.10	0.31	0.78	1.34	
Sudan	3.05	0.39	0.05	0.92	0.53	1.15	
Zaire	2.03	0.18	0.07	0.91	0.17	0.71	
Zambia	2.40	0.44	0.02	0.99	1.45	-0.49	
Zimbabwe	2.67	0.46	0.18	0.29	0.68	1.05	
Asia							
China	4.10	0.35	-0.03	2.40	0.05	1.27	0.06
India	2.50	0.28	0.05	2.34	1.31	-1.44	-0.04
Indonesia	3.80	0.14	0.18	2.60	0.37	0.46	0.05
Philippines	3.44	0.32	0.36	1.26	0.50	1.05	-0.06
Advanced Technology							
Australia	2.16	-0.15	-0.05	0.12	0.23	2.02	
France	1.66	-0.83	-0.15	0.49	0.47	1.65	
United Kingdom	1.60	-0.34	-0.11	0.11	0.36	1.58	
United States	1.85	-0.41	-0.03	-0.05	0.45	1.89	

Source: Arnade (1997).

Table : Contributions to Average Rates of Output Growth, Based on Coefficients from an Aggregate Cobb-Douglas Production Function, 1961-1987 (Percentage of Output Growth Due to Each Factor)

Country	Output	Labour	Land	Tractors	Fertilizer	Productivity	Animals
C-D coefficients		0.20	0.45	0.20	0.15		
Africa							
Kenya	2.81	23.8	4.6	11.4	38.8	21.4	
South Africa	2.14	-8.9	-4.7	14.5	36.4	62.6	
Sudan	3.05	12.8	1.6	30.2	17.4	37.8	
Zaire	2.03	8.9	3.4	44.8	8.4	35.0	
Zambia	2.40	18.3	0.8	41.3	60.4	-20.4	
Zimbabwe	2.67	17.2	6.7	10.9	25.5	39.3	
Asia							
China	4.10	8.5	-0.7	58.5	1.2	31.0	1.5
India	2.50	11.2	2.0	93.6	52.4	-57.6	-1.6
Indonesia	3.80	3.7	4.7	68.4	9.7	12.1	1.3
Philippines	3.44	9.3	10.5	36.7	14.5	30.5	-1.7
Advanced Technology							
Australia	2.16	-6.9	-2.3	5.6	10.6	93.5	
France	1.66	-50.0	-9.0	29.5	28.3	99.4	
UK	1.60	-21.3	-6.9	6.9	22.5	98.8	
United States	1.85	-22.1	-1.6	-2.7	24.3	102.2	

Source: Arnade (1997).

BOTTLENECKS TO GROWTH

The studies reviewed provide a guide to the factors that have historically affected agricultural productivity in SSA. Since conventional inputs explain most of the variation in productivity between countries in SSA, it is apparent that many of these countries still have considerable potential to raise productivity through increased use of fertilizer, machinery and livestock inputs. It has been argued that barriers to increased use of these inputs include lack of appropriate infrastructure, poor policy environments, and lack of cash to increase input purchases.

The importance of conventional inputs in SSA suggests that factors limiting their use are the most critical constraints on continued growth in agricultural productivity. Foremost among these are inadequacies in the provision of basic infrastructure, both physical and institutional. For example, limited surface transportation and communication networks in SSA increase the cost of inputs, inhibit the timely acquisition and application of inputs, and decrease access to output markets. Examples of institutional bottlenecks with similar effects include elements as diverse as political instability and constraints on access to credit and extension services. Credit market constraints are in turn driven, at least in part, by the complexities of land tenure that characterize much agricultural land in SSA. In particular, lack of individual private tenure and associated land titles as collateral may inhibit access to formal credit sources, even though customary tenure systems may offer no less security than individual private property systems.

Even more basic than concerns about physical and institutional infrastructure, however, are questions about the potential for continued increases in application of the conventional inputs that have contributed to growth in SSA agricultural productivity in the past.

For example, Crosson and Anderson (1995) report that just over one billion hectares of land are considered by FAO to be at least marginally suitable for crop production in SSA. About 213 million hectares, or just over a fifth of that, are currently in crops.

Crosson and Anderson note that if all the remaining suitable land were to be brought under crop production in the coming decades, output would increase more than enough to meet a tripling in demand by 2025, even without any increase in crop yields. Expansion on such a scale is of course unlikely, as the authors argue, because of the economic and environmental costs involved. In fact, FAO (1993) estimates that cropland area in SSA will expand by 0.9 percent per year over the next decade, which, if continued, would result in a 37 percent expansion by 2025. By contrast, the World Bank suggests that 0.5 percent annual expansion may be the maximum rate consistent with long-term sustainability.

Given the importance accorded to physical infrastructure and education as non-conventional inputs in other multi-country studies of agricultural productivity as well (Craig, Pardey and Roseboom, 1997; Antle, 1983), it is surprising that these variables have not been included in the studies exclusive to Africa. It may be that data on infrastructure are not available for a sufficiently large set of African countries. In addition, the sequence of non-conventional inputs may be important. A study of agricultural productivity in the US has shown that infrastructure investments made important contributions to agricultural productivity through the 1960s. Since that time, however, public and private R&D have become more important in spurring productivity growth in the United States. If a similar trend holds for countries where infrastructure is not yet well developed (as in much of Africa), large increases in agricultural productivity may be possible from investments in rural roads and utilities.

Other variables that deserve closer attention in studies of agricultural productivity include changes in resource quality over time and measures of political and institutional instability. Messer, Cohen and D'Costa (1998) estimated that cessation of armed conflict would have added two to five percent annually to Africa's per caput food production since 1980. Peterson's (1987) useful land quality index, which controls for irrigation, precipitation and soil nitrogen, has been used frequently in international agricultural empirical work, but provides only one (constant) number per country that fails to reflect possible changes in land quality over time. If a portion of growth in agricultural output is actually due soil fertility depletion, but soil depletion is left as an unmeasured explanatory variable, then growth in output may be incorrectly attributed to productivity growth. Based on the limited data currently available on land degradation and its productivity consequences in SSA, Crosson and Anderson (1995) estimate the average loss in agricultural productivity due to historic land degradation for Africa as a whole is about 12 percent. The authors conclude that land expansion and restoration will together contribute only about a third of the increased production necessary to meet anticipated demand in 2025. They caution that the potential for increased water supply is too limited to make a major contribution to increased production.

Crosson and Anderson argue that the remainder of the necessary production increases will have to come from adoption of a variety of more productive technologies, including improved crop varieties, increased use of fertilizer and pesticide and mechanization. Policies that will help widespread adoption of such technologies include reform of foreign exchange and tax policies that discriminate against agriculture, improvement of transportation and communications infrastructure, improved education and extension services, support for research and increased recognition of the security of property rights in land afforded by evolving local tenure systems.

Following Pingali and Heisey (1996), the technological transformation of crop production systems can be characterized in various stages, as different factors of production become scarce in succession. Pingali and Heisey describe three stages in particular with regard to cereal production, as land, labour and factors such as knowledge and management intensity become increasingly valuable. Thus, cropland expansion alone will no longer satisfy needed output growth, and further increases will need to come from intensification of production on existing cropland. Such intensification will require investment not just in basic transportation infrastructure but in the physical and institutional infrastructure necessary to improve delivery of irrigation, commercial fertilizer, extension services and other conventional and non-conventional inputs. Pingali and Heisey argue that for maize, an important food crop in much of SSA, there remains an economically exploitable gap between farmer performance and the technology frontier as represented by the yields achieved on experiment stations (in contrast to rice and wheat yields). They argue further that the technology frontier itself could be shifted more readily for maize than for rice or wheat, through transfer of technology from the more advanced countries. They caution, however, that such transfers are much less likely in SSA than in parts of Asia, where rising feed demand coincides with institutional environments that are more attractive to large private-sector seed companies.

THE ROLE OF POLICY

Agriculture in SSA is characterized by multiple constraints on accelerated productivity growth. In the absence of broad improvements in physical infrastructure, political stability and the institutional environment, the returns to any given intervention in isolation are likely to be limited as other constraints quickly become binding. In such an environment, the role of policy is twofold.

First, governments and international agencies need to invest in underlying physical and institutional infrastructure to improve the basic performance of markets by reducing the costs of transportation and transactions and by facilitating the transmission of goods, services and market signals. Improved access to fertilizer, credit and roads are among the most promising steps that could be taken along these lines. Such improvements can be expected to reduce input costs and increase access to output markets, providing both demand- and supply-side incentives for increased use of conventional inputs and output growth.

Even when markets can be structured to perform more efficiently, a second role of policy remains critical. This is the mitigation of externalities. For example, reduced fallow periods by one farmer might pose erosion problems that result in sedimentation or increased flood risk to producers downstream, or in eventual on-site resource degradation that threatens farm yields in the future. Externalities highlight the importance of well-defined institutions governing property and the distribution of costs and benefits associated with various technical and institutional innovations. They also highlight the importance of policy in influencing how these costs and benefits are distributed spatially and temporally. Hazell and Fan (1998) note the importance of investing in measures to improve productivity not only in prime agricultural areas but in less-favoured lands as well. Their results are based on analysis of Indian data; additional research is needed to determine whether similar patterns may characterize sub-Saharan Africa. An explanation for productivity growth, focusing particularly on institutional reforms that affect the performance of markets. Block (1995) found that countries that depreciated the real exchange rate tended to have higher growth rates of total factor productivity. Fulginiti and Perrin (1997) used nominal price protection as a proxy for policy reform and concluded that the countries that tax agriculture the most tend to have the most negative rates of productivity change. Fulginiti and Perrin (1993) and Hu and Antle (1993) found that an indicator of the degree of subsidization or taxation of agriculture is significant in some ranges; reducing protection would increase (decrease) productivity in countries that have been taxing (subsidizing) agriculture.

CONCLUSIONS

Agricultural production has been increasing in SSA at over two percent per year in recent years (FAO, 1999). Land used in agricultural production has increased less rapidly, resulting in an average annual increase in land productivity of 1.9 percent in SSA between 1980 and 1993 (World Bank, 1998). Labour used in agriculture has increased more rapidly than land and labour productivity declined at an average annual rate of 1.0 percent between 1980 and 1995 (World Bank, 1998). Levels of physical capital, livestock, fertilizer and non-conventional inputs have also changed, contributing to an estimated 1.3 percent annual increase in total factor productivity between 1961 and 1991.

Existing research has consistently found that about three-quarters of the variation in agricultural productivity in SSA is explained by the use of conventional inputs. This is not surprising, given the low levels of use of some conventional inputs in SSA, such as physical capital and fertilizer, relative to other developing regions and the developed countries. As a result, research suggests that there remains significant scope to improve productivity in many SSA countries through increased use of conventional inputs, particularly fertilizer, physical capital and livestock.

Based in part on such improvements in productivity, several recent research efforts have generated projections of future growth in agricultural production in SSA. FAO projects agricultural production will increase at an average annual rate of 3.0 percent between 1988-1990 and 2010, driven primarily by increases in yield (FAO, 1993; Alexandratos, 1995). The International Food Policy Research Institute projects 3.0 percent growth in grain production annually between

1990 and 2020, also due primarily to yield increases. The World Bank projects annual rates of growth in grain production of 1.5 to 2.5 percent for selected SSA countries between 2000 and 2010 (Mitchell and Ingco, 1995). The USDA's Economic Research Service (Shapouri and Rosen, 1998) projects that food production in SSA will grow at an average rate of 2.3 percent per year over the next decade through a combination of area expansion (1.3 percent per year) and yield increases (1.0 percent per year).

The ERS analysis further projects that food production in SSA would have to grow at a rate of 3.3 to 4.5 percent annually to maintain per caput consumption levels or meet nutritional requirements over the next decade. Given projected production increases on the order of 2.3 to 3.0 percent per year, these rates suggest that additional increases on the order of one to two percent per year will be required to make progress towards food security. If the World Bank's recommendation (Cleaver and Schreiber, 1994) that agricultural area expansion in SSA be limited to 0.5 percent per year on sustainability grounds is further incorporated, this gap increases to two to three percent per year.

How might such gains be realized? The sources of productivity growth in SSA agriculture indicate that continued growth of the agricultural labour force on the order of two percent per year (FAO, 1999) can be expected to increase agricultural output by about one percent per year. As land becomes increasingly scarce relative to labour, farmers will increasingly seek ways to augment land through increased application of other inputs as well. Fertilizer application rates have been declining in Africa by an average of 1.1 percent per year since 1990 (FAO, 1999); reversing this trend and increasing fertilizer use by five percent per year could increase agricultural output by an additional 0.5 percent per year. Proportionate increases in the use of machinery and in research expenditures could be expected to add similar increases to output. Expected increases in output from improved infrastructure and price policies are difficult to quantify, but such improvements are probably prerequisites to make possible the increases in productivity from the use of conventional inputs and research.

Achieving these goals poses significant challenges for policymakers. Agriculture in SSA is characterized by multiple constraints, principal among which are poverty, poor infrastructure and political instability, which combine to limit increased use of conventional inputs in SSA agriculture. Other important constraints to agricultural productivity are the quality and availability of education, research and extension services, as well as institutional uncertainties that weaken incentives to invest in the maintenance or improvement of land quality.

In the absence of broad improvements in physical infrastructure, political stability and the institutional environment, the returns to any given intervention in isolation are likely to be limited as other constraints quickly become binding. Policy reforms directed at improving physical and institutional infrastructure may not only increase use of inputs by lowering prices, but may also improve farm-gate prices of agricultural output and thus more directly stimulate output.

In addition to facilitating increased use of conventional inputs, education of the rural labour force as well as agricultural research will improve the future prospects for productivity growth in SSA. The full benefits of research are unlikely to be realized before more basic constraints are surmounted. Nevertheless continued investment in research (alongside attention to more basic sources of productivity growth) remains important due to potentially long lags in application.

6

Food Security Risk Management

At the national level, food security essentially refers to the capacity of a country to provide sufficient food for its population, and it will depend on such factors as food production, imports, food aid and within-country food distribution (in India's case, this is state-level food security). The ability to absorb shocks to such a system from drought, floods, civil strife, etc. is also relevant. At the household level, food security is here defined as access to food, that is adequate in terms of quality, quantity, safety and cultural acceptability, for all household members (Gillespie and Mason 1991). While many definitions of household security exist, they virtually all hinge on a household's ability to get sufficient food to the household door. The two main factors determining household food security - poverty and food prices - are investigated here. The cost of achieving and maintaining food security is important and can be revealed through an analysis of the proportions of either income or expenditure that is allocated to food.

DEFINING FOOD SECURITY

Food security is a flexible concept as reflected in the many attempts at definition in research and policy usage. Even a decade ago, there were about 200 definitions in published writings. Whenever the concept is introduced in the title of a study or its objectives, it is necessary to look closely to establish the explicit or implied definition. The continuing evolution of food security as an operational concept in public policy has reflected the wider recognition of the complexities of the technical and policy issues involved.

The most recent careful redefinition of food security is that negotiated in the process of international consultation leading to the World Food Summit in November 1996. The contrasting definitions of food security adopted in 1974 and 1996, along with those in official FAO and World Bank documents of the mid-1980s, are set out below with each substantive change in definition underlined. A comparison of these definitions highlights the considerable reconstruction of official thinking on food security that has occurred over 25 years. These statements also provide signposts to the policy analyses, which have reshaped our understanding of food security as a problem of international and national responsibility.

Food security as a concept originated only in the mid-1970s, in the discussions of international food problems at a time of global food crisis. The initial focus of attention was primarily on food supply problems - of assuring the availability and to some degree the price stability of basic foodstuffs at the international and national level. That supply-side, international and institutional set of concerns reflected the changing organization of the global food economy that had precipitated the crisis. A process of international negotiation followed, leading to the World Food Conference of 1974, and a new set of institutional arrangements covering information, resources for promoting food security and forums for dialogue on policy issues. The issues of famine, hunger and food crisis were also being extensively examined, following the events of the mid-1970s. The outcome was a redefinition of food security, which recognised that the behaviour of potentially vulnerable and affected people was a critical aspect.

A third, perhaps crucially important, factor in modifying views of food security was the evidence that the technical successes of the Green Revolution did not automatically and rapidly lead to dramatic reductions in poverty and levels of malnutrition. These problems were recognised as the result of lack of effective demand.

Official concepts of food security

The initial focus, reflecting the global concerns of 1974, was on the volume and stability of food supplies.

Food security was defined in the 1974 World Food Summit as:

- "Availability at all times of adequate world food supplies of basic foodstuffs to sustain a steady expansion of food consumption and to offset fluctuations in production and prices".

In 1983, FAO expanded its concept to include securing access by vulnerable people to available supplies, implying that attention should be balanced between the demand and supply side of the food security equation:

- "Ensuring that all people at all times have both physical and economic access to the basic food that they need".

In 1986, the highly influential World Bank report "Poverty and Hunger" focused on the temporal dynamics of food insecurity. It introduced the widely accepted distinction between chronic food insecurity, associated with problems of continuing or structural poverty and low incomes, and transitory food insecurity, which involved periods of intensified pressure caused by natural disasters, economic collapse or conflict.

This concept of food security is further elaborated in terms of: "Access of all people at all times to enough food for an active, healthy life".

By the mid-1990s food security was recognised as a significant concern, spanning a spectrum from the individual to the global level. However, access now involved sufficient food, indicating continuing concern with protein-energy malnutrition. But the definition was broadened to incorporate food safety and also nutritional balance, reflecting concerns about food composition and minor nutrient requirements for an active and healthy life. Food preferences, socially or culturally determined, now became a consideration. The potentially high degree of context specificity implies that the concept had both lost its simplicity and was not itself a goal, but an intermediating set of actions that contribute to an active and healthy life.

The 1994 UNDP Human Development Report promoted the construct of human security, including a number of component aspects, of which food security was only one. This concept is closely related to the human rights perspective on development that has, in turn, influenced discussions about food security.

The 1996 World Food Summit adopted a still more complex definition:

- "Food security, at the individual, household, national, regional and global levels [is achieved] when all people, at all times, have physical and economic access to sufficient, safe and nutritious food to meet their dietary needs and food preferences for an active and healthy life".

This definition is again refined in The State of Food Insecurity 2001:

- "Food security a situation that exists when all people, at all times, have physical, social and economic access to sufficient, safe and nutritious food that meets their dietary needs and food preferences for an active and healthy life".

This new emphasis on consumption, the demand side and the issues of access by vulnerable people to food, is most closely identified with the seminal study by Amartya Sen. Eschewing the use of the concept of food security, he focuses on the entitlements of individuals and households. The international community has accepted these increasingly broad statements of common goals and implied responsibilities. But its practical response has been to focus on narrower, simpler objectives around which to organise international and national public action. The declared primary objective in international development policy discourse is increasingly the reduction and elimination of poverty.

The 1996 WFS exemplified this direction of policy by making the primary objective of international action on food security halving of the number of hungry or undernourished people by 2015. Essentially, food security can be described as a phenomenon relating to individuals. It is the nutritional status of the individual household member that is the ultimate focus, and the risk of that adequate status not being achieved or becoming undermined. The latter risk describes the vulnerability of individuals in this context.

Vulnerability may occur both as a chronic and transitory phenomenon. Food security exists when all people, at all times, have physical, social and economic access to sufficient, safe and nutritious food which meets their dietary needs and food preferences for an active and healthy life. Household food security is the application of this concept to the family level, with individuals within households as the focus of concern. Food insecurity exists when people do not have adequate physical, social or economic access to food.

ESSENTIALLY OF FOOD SECURITY

At the national level, food security essentially refers to the capacity of a country to provide sufficient food for its population, and it will depend on such factors as food production, imports, food aid and within-country food distribution (in India's case, this is state-level food security). The ability to absorb shocks to such a system from drought, floods, civil strife, etc. is also relevant. At the household level, food security is here defined as access to food, that is adequate in terms of quality, quantity, safety and cultural acceptability, for all household members (Gillespie and Mason 1991). While many definitions of household security exist, they virtually all hinge on a household's ability to get sufficient food to the household door. The two main factors determining household food security - poverty and food prices - are investigated here. The cost of achieving and maintaining food security is important and can be revealed through an analysis of the proportions of either income or expenditure that is allocated to food.

Food security refers to the availability of food and one's access to it. A household is considered food-secure when its occupants do not live in hunger or fear of starvation. According to the World Resources Institute, global per capita food production has been increasing substantially for the past several decades. In 2006, MSNBC reported that globally, the number of people who are overweight has surpassed the number who are undernourished-the world had more than one billion people who were overweight, and an estimated 800 million who were undernourished. According to a 2004 article from the BBC, China, the world's most populous country, is suffering from an obesity epidemic. In India, the second-most populous country in the world, 30 million people have been added to the ranks of the hungry since the mid-1990s and 46% of children are underweight.

Worldwide around 852 million people are chronically hungry due to extreme poverty, while up to 2 billion people lack food security intermittently due to varying degrees of poverty. Six million children die of hunger every year-17,000 every day. As of late 2007, export restrictions and panic buying, US Dollar Depreciation, increased farming for use in biofuels, world oil prices at more than $100 a barrel, global population growth, climate change, loss of agricultural land to residential and industrial development, and growing consumer demand in China and India are claimed to have pushed up the price of grain. However, the role of some of these factors is under debate. Some argue the role of biofuel has been overplayed as grain prices have come down to the levels of 2006. Nonetheless, food riots have recently taken place in many countries across the world.

It is becoming increasingly difficult to maintain food security in a world beset by a confluence of "peak" phenomena, namely peak oil, peak water, peak phosphorus, peak grain and peak fish. Approximately 3.3 billion people, more than half of the planet's population, live in urban areas as of November 2007. Any disruption to farm supplies may precipitate a uniquely urban food crisis in a relatively short time. The ongoing global credit crisis has affected farm credits, despite a boom in commodity prices. Food security is a complex topic, standing at the intersection of many disciplines.

A new peer-reviewed journal of *Food Security: The Science, Sociology and Economics of Food Production and Access to Food* began publishing in 2009. In developing countries, often 70% or more of the population lives in rural areas. In that context, agricultural development among smallholder farmers and landless people provides a livelihood for people allowing them the opportunity to stay in their communities.

In many areas of the world, land ownership is not available, thus, people who want or need to farm to make a living have little incentive to improve the land.

In the US, there are approximately 2,000,000 farmers, less than 1% of the population. A direct relationship exists between food consumption levels and poverty. Families with the financial resources to escape extreme poverty rarely suffer from chronic hunger, while poor families not only suffer the most from chronic hunger, but are also the segment of the population most at risk during food shortages and famines.

Two commonly used definitions of food security come from the UN's Food and Agriculture Organization (FAO) and the United States Department of Agriculture (USDA):

- Food security exists when all people, at all times, have physical, social and economic access to sufficient, safe and nutritious food to meet their dietary needs and food preferences for an active and healthy life.
- Food security for a household means access by all members at all times to enough food for an active, healthy life. Food security includes at a minimum (1) the ready availability of nutritionally adequate and safe foods, and (2) an assured ability to acquire acceptable foods in socially acceptable ways (that is, without resorting to emergency food supplies, scavenging, stealing, or other coping strategies). (USDA)

The stages of food insecurity range from food secure situations to full-scale famine. "Famine and hunger are both rooted in food insecurity. Food insecurity can be categorized as either chronic or transitory. Chronic food insecurity translates into a high degree of vulnerability to famine and hunger; ensuring food security presupposes elimination of that vulnerability. [Chronic] hunger is not famine. It is similar to undernourishment and is related to poverty, existing mainly in poor countries."

CLIMATE CHANGE AND FOOD SECURITY

Impacts on Food Production and Availability

Climate change affects agriculture and food production in complex ways. It affects food production directly through changes in agro-ecological conditions and indirectly by affecting growth and distribution of incomes, and thus demand for agricultural produce. Impacts have been quantified in numerous studies and under various sets of assumptions. A selection of these results is presented in *Quantifying the Impacts on Food Security*. Here it is useful to summarize the main alterations in the agro-ecological environment that are associated with climate change.

Changes in temperature and precipitation associated with continued emissions of greenhouse gases will bring changes in land suitability and crop yields. In particular, the Intergovernmental Panel on Climate Change (IPCC) considers four families of socio-economic development and associated emission scenarios, known as Special Report on Emissions Scenarios (SRES) A2, B2, A1, and B1. Of relevance to this review, of the SRES scenarios, A1, the "business-as-usual scenario," corresponds to the highest emissions, and B1 corresponds to the lowest.

The other scenarios are intermediate between these two. Importantly for agriculture and world food supply, SRES A2 assumes the highest projected population growth of the four (United Nations high projection) and is thus associated to the highest food demand. Depending on the SRES emission scenario and climate models considered, global mean surface temperature is projected to rise in a range from 1.8°C (with a range from 1.1°C to 2.9°C for SRES B1) to 4.0°C (with a range from 2.4°C to 6.4°C for A1) by 2100. In temperate latitudes, higher temperatures are expected to bring predominantly benefits to agriculture: the areas potentially suitable for cropping will expand, the length of the growing period will increase, and crop yields may rise.

A moderate incremental warming in some humid and temperate grasslands may increase pasture productivity and reduce the need for housing and for compound feed. These gains have to be set against an increased frequency of extreme events, for instance, heat waves and droughts in the Mediterranean region or increased heavy precipitation events and flooding in temperate regions, including the possibility of increased coastal storms; they also have to be set against the fact that semiarid and arid pastures are likely to see reduced livestock productivity and increased livestock mortality.

In drier areas, climate models predict increased evapotranspiration and lower soil moisture levels. As a result, some cultivated areas may become unsuitable for cropping and some tropical grassland may become increasingly arid. Temperature rise will also expand the range of many agricultural pests and increase the ability of pest populations to survive the winter and attack spring crops.

Another important change for agriculture is the increase in atmospheric carbon dioxide (CO_2) concentrations. Depending on the SRES emission scenario, the atmospheric CO_2 concentration is projected to increase from H"379 ppm today to >550 ppm by 2100 in SRES B1 to >800 ppm in SRES A1FI. Higher CO_2 concentrations will have a positive effect on many crops, enhancing biomass accumulation and final yield. However, the magnitude of this effect is less clear, with important differences depending on management type (*e.g.*, irrigation and fertilization regimes) and crop type. Experimental yield response to elevated CO_2 show that under optimal growth conditions, crop yields increase at 550 ppm CO_2 in the range of 10 per cent to 20 per cent for C_3 crops (such as wheat, rice, and soybean), and only 0–10 per cent for C_4 crops such as maize and sorghum. Yet the nutritional quality of agricultural produce may not increase in line with higher yields. Some cereal and forage crops, for example, show lower protein concentrations under elevated CO_2 conditions.

Finally, a number of recent studies have estimated the likely changes in land suitability, potential yields, and agricultural production on the current suite of crops and cultivars available today. Therefore, these estimates implicitly include adaptation using available management techniques and crops, but excluding new cultivars from breeding or biotechnology. These studies are in essence based on the FAO/International Institute for Applied Systems Analysis (IIASA) agro-ecological zone (AEZ) methodology. For instance, pioneering work in ref. 9 suggested that total land and total prime land would remain virtually unchanged at the current levels of 2,600 million and 2,000 million hectares (ha), respectively. The same study also showed, however, more pronounced regional shifts, with a considerable increase in suitable cropland at higher latitudes (developed countries +160 million ha) and a corresponding decline of potential cropland at lower latitudes (developing countries "110 million ha). An even more pronounced shift within the quality of cropland is predicted in developing countries. The net decline of 110 million ha is the result of a massive decline in agricultural prime land of H"135 million ha, which is offset by an increase in moderately suitable land of >20 million ha. This quality shift is also reflected in the shift in land suitable for multiple cropping. In sub-Saharan Africa alone, land for double cropping would decline by between 10 million and 20 million ha, and land suitable for triple copping would decline by 5 million to 10 million ha. At a regional level, similar approaches indicate that under climate change, the biggest losses in suitable cropland are likely to be in Africa, whereas the largest expansion of suitable cropland is in the Russian Federation and Central Asia.

Impacts on the Stability of Food Supplies

Global and regional weather conditions are also expected to become more variable than at present, with increases in the frequency and severity of extreme events such as cyclones, floods, hailstorms, and droughts. By bringing greater fluctuations in crop yields and local food supplies and higher risks of landslides and erosion damage, they can adversely affect the stability of food supplies and thus food security.

Neither climate change nor short-term climate variability and associated adaptation are new phenomena in agriculture, of course. As shown, for instance, in ref. 9, some important agricultural areas of the world like the Midwest of the United States, the northeast of Argentina, southern Africa, or southeast Australia have traditionally experienced higher climate variability than other regions such as central Africa or Europe. They also show that the extent of short-term fluctuations has changed over longer periods of time. In developed countries, for instance, short-term climate variability increased from 1931 to 1960 as compared with 1901 to 1930, but decreased strongly in the period from 1961 to 1990. What is new, however, is the fact that the areas subject to high climate variability are likely to expand, whereas the extent of short-term climate variability is likely to increase across all regions. Furthermore, the rates and levels of projected warming may exceed in some regions the historical experience.

If climate fluctuations become more pronounced and more widespread, droughts and floods, the dominant causes of short-term fluctuations in food production in semiarid and subhumid areas, will become more severe and more frequent. In semiarid areas, droughts can dramatically reduce crop yields and livestock numbers and productivity.

Again, most of this land is in sub-Saharan Africa and parts of South Asia, meaning that the poorest regions with the highest level of chronic undernourishment will also be exposed to the highest degree of instability in food production. How strongly these impacts will be felt will crucially depend on whether such fluctuations can be countered by investments in irrigation, better storage facilities, or higher food imports. In addition, a policy environment that fosters freer trade and promotes investments in transportation, communications, and irrigation infrastructure can help address these challenges early on.

Impacts of Climate Change on Food Utilization

Climate change will also affect the ability of individuals to use food effectively by altering the conditions for food safety and changing the disease pressure from vector, water, and food-borne diseases. The IPPC Working Group II provides a detailed account of the health impacts of climate change of its fourth assessment report. It examines how the various forms of diseases, including vector-borne diseases such as malaria, are likely to spread or recede with climate change. This chapter focuses on a narrow selection of diseases that affect food safety directly, *i.e.*, food and water-borne diseases.

The main concern about climate change and food security is that changing climatic conditions can initiate a vicious circle where infectious disease causes or compounds hunger, which, in turn, makes the affected populations more susceptible to infectious disease. The result can be a substantial decline in labour productivity and an increase in poverty and even mortality. Essentially all manifestations of climate change, be they drought, higher temperatures, or heavy rainfalls have an impact on the disease pressure, and there is growing evidence that these changes affect food safety and food security.

The recent IPCC report also emphasizes that increases in daily temperatures will increase the frequency of food poisoning, particularly in temperate regions. Warmer seas may contribute to increased cases of human shellfish and reef-fish poisoning (ciguatera) in tropical regions and a poleward expansion of the disease. However, there is little new evidence that climate change significantly alters the prevalence of these diseases. Several studies have confirmed and quantified the effects of temperature on common forms of food poisoning, such as salmonellosis. These studies show an approximately linear increase in reported cases for each degree increase in weekly temperature. Moreover, there is evidence that temperature variability affects the incidence of diarrhoeal disease. A number of studies found that rising temperatures were strongly associated with increased episodes of diarrhoeal disease in adults and children. These findings have been corroborated by analyses based on monthly temperature observations. Several studies report a strong correlation between monthly temperature and diarrhoeal episodes on the Pacific Islands, Australia, and Israel.

Extreme rainfall events can increase the risk of outbreaks of water-borne diseases particularly where traditional water management systems are insufficient to handle the new extremes. Likewise, the impacts of flooding will be felt most strongly in environmentally degraded areas, and where basic public infrastructure, including sanitation and hygiene, is lacking. This will raise the number of people exposed to water-borne diseases (*e.g.*, cholera) and thus lower their capacity to effectively use food.

Impacts of Climate Change on Access to Food

Access to food refers to the ability of individuals, communities, and countries to purchase sufficient quantities and qualities of food. Over the last 30 years, falling real prices for food and rising real incomes have led to substantial improvements in access to food in many developing countries. Increased purchasing power has allowed a growing number of people to purchase not only more food but also more nutritious food with more protein, micronutrients, and vitamins. East Asia and to a lesser extent the Near-East/ North African region have particularly benefited from a combination of lower real food prices and robust income growth. From 1970 to 2001, the prevalence of hunger in these regions, as measured by FAO's indicator of undernourishment, has declined from 24 per cent to 10.1 per cent and 44 per cent to 10.2 per cent respectively. In East Asia, it was endogenous income growth that provided the basis for the boost in demand for food, which was largely produced in the region; in the Near-East North African region demand was spurred by exogenous revenues from oil and gas exports, and additional food supply came largely from imports. But in both regions, improvements in access to food have been crucial in reducing hunger and malnutrition.

FAO's longer-term outlook to 2050 suggests that the importance of improved demand side conditions will even become more important over the next 50 years. The regions that will see the strongest reductions in the prevalence of undernourishment are those that are expected to see the highest rates of income growth. South Asia stands to benefit the most. Spurred by high income growth the region is expected to reduce the prevalence of undernourishment from >22 per cent currently to 12 per cent by 2015 and just 4 per cent by 2050. Progress is also expected for sub-Saharan Africa, but improvements will be less pronounced and are expected to set in later. Over the next 15 years, for instance, the prevalence of undernourishment will decline less than in other regions, from H-33 per cent to a still worrisome 21 per cent, as significant constraints (soil nutrients, water, infrastructure, etc.) will limit the ability to further increase food production locally, while continuing low levels of income rule out the option of importing food. In the long run, however, sub-Saharan Africa is expected to see a more substantial decline in hunger; by 2050, <6 per cent of its total population are expected to suffer from chronic hunger. It is important to note that these FAO projections do not take into account the effects of climate change.

By coupling agro-ecologic and economic models, others have gauged the impact of climate change on agricultural gross domestic product (GDP) and prices. At the global level, the impacts of climate change are likely to be very small; under a range of SRES and associated climate-change scenarios, the estimates range from a decline of "1.5 per cent to an increase of +2.6 per cent by 2080.

At the regional level, the importance of agriculture as a source of income can be much more important. In these regions, the economic output from agriculture itself will be an important contributor to food security. The strongest impact of climate change on the economic output of agriculture is expected for sub-Saharan Africa, which means that the poorest and already most food-insecure region is also expected to suffer the largest contraction of agricultural incomes. For the region, the losses in agricultural GDP, compared with no climate change, range from 2 per cent to 8 per cent for the Hadley Centre Coupled Model, version 3 and 7–9 per cent for the Commonwealth Scientific and Industrial Research Organisation projections.

Impacts on Food Prices

Essentially all SRES development paths describe a world of robust economic growth and rapidly shrinking importance of agriculture in the long run and thus a continuation of a trend that has been underway for decades in many developing regions. SRES scenarios describe a world where income growth will allow the largest part of the world's population to address possible local production shortfalls through imports and, at the same time, find ways to cope with safety and stability issues of food supplies. It is also a world where real incomes rise more rapidly than real food prices, which suggests that the share of income spent on food should decline and that even high food prices are unlikely to create a major dent in the food expenditures of the poor. However, not all parts of the world perform equally well in the various development paths and not all development paths are equally benign for growth. Where income levels are low and shares of food expenditures are high, higher prices for food may still create or exacerbate a possible food security problem.

There are a number of studies that have ventured to measure the likely impacts of climate change on food prices. The basic messages that emerge from these studies are: first, on average, prices for food are expected to rise moderately in line with moderate increases of temperature (until 2050); some studies even foresee a mild decline in real prices until 2050. Second, after 2050 and with further increases in temperatures, prices are expected to increase more substantially. In some studies and for some commodities (rice and sugar) prices are forecast to increase by as much as 80 per cent above their reference levels without climate change. Third, price changes expected from the effects of global warming are, on average, much smaller than price changes from socio-economic development paths.

For instance, the SRES A2 scenario would imply a price increase in real cereal prices by H-170 per cent. The (additional) price increase caused by climate change (in the Hadley Centre Coupled Model, version 3, climate change case) would only be 14.4 per cent. Overall, this appears to be the sharpest price increase reported and it is not surprising that this scenario would imply a stubbornly high number of undernourished people until 2080. However, it is also needless to say that a constant absolute number of undernourished people would still imply a sharp decline in the prevalence of hunger, and, given the high population assumptions in the SRES A2 world (>13.6 billion people globally and >11.6 billion in the developing world) this would imply a particularly sharp drop in the prevalence from currently 17 per cent to H-7 per cent by 2080.

NATIONAL FOOD SECURITY

Agriculture

Agriculture remains by far the most important area of the India economy, accounting for about 63% of the labour force in 1985-1987, although its importance has decreased over the years. In 1989 it accounted for 28.2% of GDP, compared to 42.8% in 1973, indicating low agricultural labour productivity. About two-thirds of the crops cultivated for domestic markets are grains, mostly rice, wheat, sorghum and maize.

In India cropping activities are year-round provided water is available for crops. In Northern India there are two distinct crops seasons - the *kharif* (July to October), and the *rabi* (October to March). In some parts of the country there are no such distinct seasons. Based on the crops, the country can be divided into the following agricultural regions.

(i) The rice region including a very large part of the north-east and south-east of India, with another strip along the western coast.

(ii) The wheat region, occupying most of the northern, western and central India.

(iii) The millet-sorghum region, comprising Rajasthan, Madhya Pradesh and the Deccan Plateau in the centre of the Indian Peninsula.

(iv) The temperate Himalayan regions of Kashmir, Himachal Pradesh and Uttar Pradesh where potatoes are as important as cash crops and tree fruits form a large part of agricultural production.

(v) The plantation crop regions of Assam (tea) and South Indian hill regions (coffee).

Rice is grown in high rainfall areas or where supplemental irrigation is available to ensure good yields. In southern states, namely Andhra Pradesh, Tamil Nadu and Kerala, rice is grown in more than one season and mostly under irrigation or under sufficient rainfall. Together these three states have over 6.0 million hectares representing over 17% of the all-India area under rice. Important alternative crops in Andhra Pradesh are pulses, groundnut, millets, maize, sugarcane and tobacco. In Kamataka, alternative crops to rice are *ragi,* millets, cotton etc. In Maharashtra rice is grown mostly in the Konkan area, over 1.3 million hectares, along with *ragi,* pulses, oil seeds, groundnut etc. In the states of Gujarat, Jammu and Kashmir, Rajasthan and Himachal Pradesh, rice is a minor crop. Rice is grown on more than 50% of cropped area in Orissa and on 4.3 million hectares (11.7% of the all India rice area) in Madhya Pradesh. In West Bengal and Uttar Pradesh rice is grown on 14.2 and 12.4% respectively of the all-India area.

Generally wheat and gram are concentrated in the sub-tropical region in northern India, where they can substitute for each other. The core of the wheat region responsible for 70% of the area and 76% of production comprises Punjab, Haryana, Delhi, Uttar Pradesh and Gujarat in the west and Bihar and West Bengal in the east.

With the increase in population, irrigated area is increasing, and as agricultural science advances, most of the extensive *cropping* patterns are giving way to intensive cropping. The development in minor irrigation works has provided farmers with opportunities to crop their land all year around with high-yielding varieties. This intensive cropping requires an easy and ready availability of balanced fertilizers and plant protection chemicals and an appropriate price policy for inputs.

Food production

In the early 1970s, a global food crisis, oil price hikes, unfavourable weather conditions, droughts (especially in Western India) and severe floods (during 1974 in the North East) all combined to threaten the two pillars of India's food strategy - a political commitment to self-reliance in food and a determination to prevent famine. Ultimately, however, these beliefs were reinforced by these challenges.

The 'green revolution' package of high-yielding crop varieties, irrigation and complementary inputs, which enabled farmers to increase per acre yields, has helped agriculture keep pace with population growth, and led to national self-sufficiency (see below).

At the national level to the present though, India has made rapid strides in agricultural production, with food production increasing from 108.4 million tonnes in 1970-71 to a record production of 176.5 million tonnes in 1990-91 (GOI 1991), though this has been offset by an increase in population from 551 million to 844 million during this period. The per capita production of wheat in particular, and to a lesser extent rice, increased most significantly, as the use of inputs, and the area irrigated expanded. The food production per capita index, with 1971-80 as base (= 100), indicates an increase early in the 1980s followed by a levelling off in the mid-1980s, a drop in the drought years 1986-87, then a marked increase to 1989, before dropping again in 1990.

In the early 1990s, however, there are signs that such a strategy is running into diminishing returns, as a result of decreasing fertiliser productivities and long-term extraction of soil minerals. The rate of growth of foodgrain production is said to be decelerating (Sarma and Gandhi 1990), raising the question of sustainability of growth and possible exhaustion of the green revolution potential. Furthermore, although evidence has been presented to show that the technologies involved (at least since the mid-1980s) are largely scale-neutral (Hazell and Ramaswamy 1991), despite an early bias towards richer farmers, they are still only effective if a constant supply of water and fertilizer can be assured.

In the (usually poorer) villages dependent on rainfed coarse grains - such as maize, sorghum and other millets - there has been no 'revolution' and here lies the challenge for the future. Initial biases towards wheat and rice and towards certain states, e.g. the Punjab and Haryana, need to be balanced, and ways sought to promote agricultural development in poorer rainfed areas.

The regional development of food production has been uneven. To some extent this simply reflects differences in the crop composition and seasonality of agriculture.

There have also been regional variations in performance for individual crops which have accentuated existing regional disparities. Initial gains accrued particularly to the north-western states of Punjab, Haryana and the western half of Uttar Pradesh, which were environmentally more favourable and better equipped to give high returns to the new wheat technology package.

Irrigated areas of Tamil Nadu were also able to exploit the new rice technology. Much poorer performance has been experienced in the states of eastern India with predominantly rain-fed agriculture and in the dryland agriculture of the states of Andhra Pradesh, Kamataka, Maharashtra and Madhya Pradesh - states characterized by a history, of instability in agricultural production and by a high incidence of poverty. Other aspects of agrarian structure such as gross inequities of land distribution with high degrees of landlessness, more or less disguised sharecropping and other forms of tenancy and exploitation through the marketing system as well as poorly developed irrigation, drainage and flood protection and a low uptake of fertilizer may explain the particularly dismal record of rice yields in the Eastern region.

Dryland agriculture is characterized by high female participation rates, very low and highly variable yields, a limited generation and heavy concentration of control over investible surplus, generalized lack of creditworthiness, small and sporadic creation of marketed surpluses and the distress commercialization (under conditions of debt and relatively low prices) of the agriculture of small and marginal farmers. Little progress has been made in developing strains of seed with high moisture stress resistance which would be suitable for use in dryland areas. Indeed, until recently there has been a bias in research funding away from these crops.

Drought and food distribution

The 1986-87 drought was the worst since before Independence, driving the agricultural strategy off-course, and necessitating the importing of cereals on a significant scale in 1988-89 (even though the drought broke in mid-1988). The precise definition of drought is difficult to provide. As Healthcote (1973) put it, "There are probably as many definitions of drought as there are uses of water". In 1987, India experienced what was labelled as the country's worst drought both in terms of intensity and geographical spread - 252 districts out of 412 in the country received deficient rainfall. During 1986-87, food production dropped to 143 million tonnes and then in 1987-88 to 138.4 million tonnes - see trend in Figure 10 (GOI 1991).

The adverse impact of the drought on the country's economy in terms of the drop in agricultural production, industrial output, reduced purchasing power and the concomitant rise in unemployment, particularly of the rural labour force, was considerable. The results of a diet and nutrition survey during the 1987 drought however showed that widespread hunger and its consequences (experienced earlier in the 1960s and 1970s) were averted in 1987.

This was achieved through the wide geographical distribution of grain buffer stocks within the Public Distribution System (PDS).. These buffer stocks had become the symbol of self-reliance during the seventies and eighties, as a food security net for the population. The Public Distribution System (PDS) involves the distribution of subsidised foodgrains via a network of fair-price shops to the eligible poor with ration cards. Figure 11 shows how public distribution rose steadily from 1975, until buffer stocks were run down during the drought of 1987-88 (to the point where imports were necessary again in 1988). While buffers stocks may be costly to maintain, the 1987-88 drought demonstrated that such a food security strategy is cheaper than importing food on a large scale. States differ as regards the coverage, efficiency and effectiveness of the PDS, with states with a better infrastructure (e.g. Punjab and Haryana) - and urban areas *per se*.- often being better served, despite their lower relative need vis a vis poorer States such as Bihar, Orissa and Uttar Pradesh.

India has not experienced famine on any significant scale since Independence, a period including the worst drought this century (1986-88) for wide areas of western and central India. This is historically and comparatively an impressive achievement. Post-Independence policy was formed in the aftermath of the Great Bengal Famine. The dramatic measures to prevent famine in Bihar in 1966/67, to contain the food crisis in 1973/74, as well as to feed 10-12 million Bangladeshi refugees in 1971 indicate the overwhelming priority of famine prevention. After the so-called "Bihar Famine" of 1966/67, this has also been accomplished without significant direct external assistance. The prevention of famine in India now has a political importance that in the short run will also outweigh any other consideration. There is therefore a battery of famine warning and prevention measures, representing a continuation and strengthening of earlier Famine Code practice (Clay *et al.* 1988).

An important qualification to this success, however, is establishing and maintaining a rudimentary food security net for both rural and urban poor throughout India, has also been accomplished without significant inroads into the massive continuing problem of structural poverty.

Per capita food availability

The per capita availability of food is a function of population and food production with adjustments made for exports, imports, wastage, livestock feed, seeds and stocks at the beginning of the year.

Cereals. The per capita net daily availability of cereals has not varied significantly in the last 20 years, being 417g in 1970-71 and 438 g in 1989-90 (GOI 1991). From the early 1950s to the early 1980s, per capita foodgrain availability fluctuated within a narrow band of 150-170 kgs per year, with a statistically insignificant trend.

Thus, foodgrain production growth seems to have been absorbed by population growth, reduction in imports and increase in stocks (Sarma and Gandhi 1990).

Pulses. Pulses account for less than 8% of total food grain production in the country. The pulse production has not kept pace with population growth. Per capita availability declined from 51.2g in 1970-71 to 33.1g in 1987-88 before rising to 36.5g in 1989-90 (GOI 1991).

This was not only due to a reduction in land under pulse cultivation but also to a decline in yield per hectare - there has been no breakthrough in technology as with wheat and rice.

Based on the 1988 population, body weights and physical activity pattern, the daily per caput energy requirement has been estimated to be 2200 kcals. (Reddy 1990). After considering post-harvest and distributional losses, the per caput requirement of foodgrain would come to 605 g/day with the total food grain requirement at production level being 178 million tonnes.

The food grain production in 1990-91 is 176.5 million tonnes, close to the estimated requirement indicating virtual *national-level* self sufficiency in this respect at present.

HOUSEHOLD FOOD SECURITY

Agriculture development and food production are central to national food security. In addition, there are several other factors that determine the food security at household level. These include food prices, food availability, employment and purchasing power, government subsidies and the food rationing system.

RISKS TO FOOD SECURITY

Fossil Fuel Dependence

While agricultural output increased as a result of the Green Revolution, the energy input into the process (that is, the energy that must be expended to produce a crop) has also increased at a greater rate, so that the ratio of crops produced to energy input has decreased over time. Green Revolution techniques also heavily rely on chemical fertilizers, pesticides and herbicides, some of which must be developed from fossil fuels, making agriculture increasingly reliant on petroleum products.

Between 1950 and 1984, as the Green Revolution transformed agriculture around the globe, world grain production increased by 250%. The energy for the Green Revolution was provided by fossil fuels in the form of fertilizers (natural gas), pesticides (oil), and hydrocarbon fuelled irrigation.

David Pimentel, professor of ecology and agriculture at Cornell University, and Mario Giampietro, senior researcher at the National Research Institute on Food and Nutrition (INRAN), place in their study *Food, Land, Population and the U.S. Economy* the maximum U.S. population for a sustainable economy at 200 million. To achieve a sustainable economy and avert disaster, the United States must reduce its population by at least one-third, and world population will have to be reduced by two-thirds, says the study.

The authors of this study believe that the mentioned agricultural crisis will only begin to impact us after 2020, and will not become critical until 2050.

The oncoming peaking of global oil production (and subsequent decline of production), along with the peak of North American natural gas production will very likely precipitate this agricultural crisis much sooner than expected. Geologist Dale Allen Pfeiffer claims that coming decades could see spiraling food prices without relief and massive starvation on a global level such as never experienced before.

However, one should take note that, (numbers taken from the CIA World Factbook), the country of Bangladesh achieved food self-sufficiency in 2002 with both a far higher population density than the USA (~1000 inhabitants per square kilometer in comparison to just 30/km for the USA-so this is more than 30 times as many), and at only a tiny fraction of the USA's usage of oil, gas, and electricity.

Also, pre-industrial Chinese mini-farmers/gardeners developed techniques to feed a population of more than 1000 people per square kilometer. Hence, the dominant problem is not energy availability but the need to stop and revert soil degradation. Hybridization, genetic engineering and loss of biodiversity

In agriculture and animal husbandry, the Green Revolution popularized the use of conventional hybridization to increase yield by creating "high-yielding varieties". Often the handful of hybridized breeds originated in developed countries and were further hybridized with local varieties in the rest of the developing world to create high yield strains resistant to local climate and diseases.

Local governments and industry have been pushing hybridization which has resulted in several of the indigenous breeds becoming extinct or threatened. Disuse because of unprofitability and uncontrolled intentional and unintentional cross-pollination and crossbreeding (genetic pollution), formerly huge gene pools of various wild and indigenous breeds have collapsed causing widespread genetic erosion and genetic pollution. This has resulted in loss of genetic diversity and biodiversity as a whole.

A genetically modified organism (GMO) is an organism whose genetic material has been altered using the genetic engineering techniques generally known as recombinant DNA technology. Genetically Modified (GM) crops today have become a common source for genetic pollution, not only of wild varieties but also of other domesticated varieties derived from relatively natural hybridization.

Genetic erosion coupled with genetic pollution may be destroying unique genotypes, thereby creating a hidden crisis which could result in a severe threat to our food security. Diverse genetic material could cease to exist which would impact our ability to further hybridize food crops and livestock against more resistant diseases and climatic changes.

Genetic Erosion in Agricultural and Livestock Biodiversity

Genetic erosion in agricultural and livestock biodiversity is the loss of genetic diversity, including the loss of individual genes, and the loss of particular combinants of genes (or gene complexes) such as those manifested in locally adapted landraces of domesticated animals or plants adapted to the natural environment in which they originated. The term genetic erosion is sometimes used in a narrow sense, such as for the loss of alleles or genes, as well as more broadly, referring to the loss of varieties or even species.

The major driving forces behind genetic erosion in crops are: variety replacement, land clearing, overexploitation of species, population pressure, environmental degradation, overgrazing, policy and changing agricultural systems. The main factor, however, is the replacement of local varieties of domestic plants and animals by high yielding or exotic varieties or species. A large number of varieties can also often be dramatically reduced when commercial varieties (including GMOs) are introduced into traditional farming systems. Many researchers believe that the main problem related to agro-ecosystem management is the general tendency towards genetic and ecological uniformity imposed by the development of modern agriculture.

Intellectual Property Rights

There is much debate on whether IPRs hurt or harm independent development in terms or agriculture and food production. Hartmut Meyer and Annette von Lossau describe both sides of the issue, while saying "Among scholars, the thesis that the impetus to self-determined development and the protection of intellectual property go hand in hand is disputed-to put it mildly.

Many studies have concluded that there is virtually no positive correlation between establishing self-sustained economic growth and ensuring protection of intellectual property rights.

Price Setting

On April 30, 2008 Thailand announces the project of the creation of the Organisation of Rice Exporting Countries with the potential to develop into a price-fixing cartel for rice.

Treating food the same as other internationally traded commodities. On October 23, 2008, Associated Press reported the following: "Former President Clinton told a U.N. gathering Thursday [Oct 16, 2008] that the global food crisis shows "we all blew it, including me," by treating food crops "like colour TVs" instead of as a vital commodity for the world's poor....Clinton criticized decades of policymaking by the World Bank, the International Monetary Fund and others, encouraged by the U.S., that pressured Africans in particular into dropping government subsidies for fertilizer, improved seed and other farm inputs as a requirement to get aid. Africa's food self-sufficiency declined and food imports rose. Now skyrocketing prices in the international grain trade—on average more than doubling between 2006 and early 2008—have pushed many in poor countries deeper into poverty."

QUANTIFYING THE IMPACTS ON FOOD SECURITY

A number of studies have recently quantified the impacts of climate change on food security. In terms of quantifying agronomic yield change projections, these studies are either based on the AEZ tools developed by the IIASA analysis or the Decision Support System for Agrotechnology Transfer suite of crop models; all use the IIASA-Basic Linked System (BLS) economic model for assessing economic impacts. These tools, with some modifications relating to how crop yield changes are simulated, have also been used by others to undertake similar assessments and provide sensitivity analyses across a range of SRES and general circulation model (GCM) projections.

Many other simulations have also examined the effects of climate change with and without adaptation (induced technological progress, domestic policy change, international trade liberalization, etc.), with and without mitigation (*e.g.*, CO_2 stabilization, variants for temperature, rainfall change and distribution) or provide impact assessments for different speeds of climate change. The quantitative results for food security, trying to illuminate some of the differences and extract the main messages that emerge from the various studies. Unless indicated, all simulation results discussed below include the combined effects of climate change and elevated CO_2 on crops. The key messages can be summarized as follows:

First, it is very likely that climate change is likely to increase the number of people at risk of hunger compared with reference scenarios with no climate change; the exact impacts will, however, strongly depend on the projected socio-economic developments. For instance, it is estimated that climate change would increase the number of undernourished in 2080 by 5–26 per cent, compared with no climate change or by between 5 million and 10 million (B1 SRES) and 120 million to 170 million people (A2 SRES), with within-SRES ranges depending on GCM climate projections. Using only one GCM scenario, others projected small reductions by 2080, *i.e.*, –5 per cent, or –10 million (B1) to –30 million (A2) people, and slight increases of +13 per cent–26 per cent, or H–10 million (B2) to 30 million (A1) people.

Second, it is likely that the magnitude of these climate impacts will be small compared with the impact of socio-economic development. The limitations of socio-economic forecasts, these studies suggest that robust economic growth and a decline in population growth projected for the 21st century will, in all but one scenario (SRES A2), significantly reduce the number of people at risk of hunger in 2080. At any rate, the prevalence of undernourishment will decline as all scenarios assume that world population will continue to grow to 2080, albeit at lower rates.

Compared with FAO estimates of 820 million undernourished in developing countries today, several studies estimate reductions of >75 per cent by 2080, or H–560 million to 700 million people, projecting 100 million to 240 million undernourished by 2080 (A1, B1, and B2). As mentioned, the only exception is scenario A2, where the number of the hungry is forecast to decrease only slightly by 2080; but the higher population growth rates in A2 compared with other scenarios mean that the prevalence of undernourishment will decline drastically. However, these analyses also confirm that the progress will be unevenly distributed over the developing world, and more importantly progress will be slow during the first decades of the outlook. With or without climate change, the millennium development goal of halving the prevalence of hunger by 2015 is unlikely to be realized before 2020–2030.

In addition to socio-economic pressures considered within the IPCC SRES scenarios, food production may increasingly compete with bio-energy in coming decades; studies addressing possible consequences for world food supply have only started to surface, providing both positive and negative views. Importantly, none of the major world food models discussed herein have yet considered such competition.

Third, sub-Saharan Africa is likely to surpass Asia as the most food-insecure region. However, this is largely independent of climate change and is mostly the result of the socio-economic development paths assumed for the different developing regions in the SRES scenarios. Throughout most SRES and climate-change scenarios sub-Saharan Africa accounts for 40–50 per cent of global hunger by 2080, compared with H–24 per cent today; in some simulations sub-Saharan Africa even accounts for 70–75 per cent of global undernourishment by 2080. These high estimates have emerged from slower growth variants of the A2 and B2 scenarios; also an A2 variant with slower population growth yields a sharper concentration of hunger in sub-Saharan Africa. For regions other than sub-Saharan Africa, results largely depend on GCM scenarios and therefore are highly uncertain.

Fourth, although there is significant uncertainty on the effects of elevated CO_2 on crop yields, this uncertainty is carried to a much lesser extent on food security. This result emerges from a comparison of climate change simulations with and without CO_2 fertilization effects on crop yields. As can be seen, higher CO_2 fertilization would not greatly affect global projections of hunger. In view of the fact that essentially all SRES worlds are characterized by much higher real incomes, much improved transportation and communication options, and still sufficient global food production, the somewhat smaller supplies will not be able to make a dent in global food security outcomes.

Many studies find that climate change without CO_2 fertilization would reduce the number of undernourished people by 2080 only by some 20 million to 140 million (120 million to 380 million for SRES A1, B1, and B2 without, compared with 100 million to 240 million with CO_2 fertilization effect). The exception again in these studies is SRES A2, under which the assumption of no CO_2 fertilization results in a projected range of 950 million to 1,300 million people undernourished in 2080, compared with 740 million to 850 million projected under climate change, but with CO_2 effects on crops.

Finally, recent research suggests large positive effects of climate stabilization for the agricultural sector. However, as the stabilizing effects of mitigation measures can take several decades to be realized from the moment of implementation, the benefits for crop production may be realized only in the second half of this century. Importantly, even in the presence of robust global long-term benefits, regional and temporal patterns of winners and losers that can be projected with current tools are highly uncertain and depend critically on the underlying GCM projections.

Uncertainties and Limitations

The finding that socio-economic development paths have an important bearing on future food security and that they are likely to top the effects of climate change should not, or at least not only, be interpreted as a probability-based forecast. This is because SRES scenarios offer a range of possible outcomes "without any sense of likelihood". Yet SRES scenarios, like all scenarios, do not overcome the inability to accurately project future changes in economic activity, emissions, and climate.

Second, the existing global assessments of climate change and food security have only been able to focus on the impacts on food availability and access to food, without quantification of the likely important climate change effects on food safety and vulnerability (stability).

This means that these assessments neither include potential problems arising from additional impacts due to extreme events such as drought and floods nor do they quantify the potential impacts of changes in the prevalence of food-borne diseases (positive as well as negative) or the interaction of nutrition and health effects through changes in the proliferation of vector-borne diseases such as malaria. On the food availability side, they also exclude the impacts of a possible sea-level rise for agricultural production or those that are associated with possible reductions of marine or freshwater fish production.

Third, it is important to note that even in terms of food availability, all current assessments of world food supply have focused only on the impacts of mean climate change, *i.e.*, they have not considered the possibility of significant shifts in the frequency of extreme events on regional production potential, nor have they considered scenarios of abrupt climate or socio-economic change; any of these scenario variants is likely to increase the already negative projected impacts of climate change on world food supply. Models that take into account the specific biophysical, technological, and market responses necessary to simulate realistic adaptation to such events are not yet available. Fourth, this review finds that recent global assessments of climate change and food security rest essentially on a single modelling framework, the IIASA system, which combines the FAO/IIASA AEZ model with various GCM models and the IIASA BLS system, or on close variants of the IIASA system.

This has important implications for uncertainty, given that the robustness of all these assessments strongly depends on the performance of the underlying models. There is, therefore, a clear need for continued and enhanced validation efforts of both the agro-climatological and food trade tools developed at IIASA and widely used in the literature. Finally, we note that assessments that do not only provide scenarios, but also attach probabilities for particular outcomes to come true could provide an important element for improved or at least better-informed policy decisions. A number of possibilities are offered to address the related modelling challenges. One option would be to produce such estimates with probability-based estimates of the (key) model parameters. Alternatively, the various scenarios could be constructed so that they reflect expert judgements on a particular issue. It would be desirable to attach probabilities to existing scenarios. Information on how likely the suggested outcomes are would contribute greatly to their usefulness for policy makers and help justify (or otherwise) policy measures to adapt to or mitigate the impacts of climate change on food security.

Climate change will affect all four dimensions of food security, namely food availability (*i.e.*, production and trade), access to food, stability of food supplies, and food utilization. The importance of the various dimensions and the overall impact of climate change on food security will differ across regions and over time and, most importantly, will depend on the overall socio-economic status that a country has accomplished as the effects of climate change set in.

Essentially all quantitative assessments show that climate change will adversely affect food security. Climate change will increase the dependency of developing countries on imports and accentuate existing focus of food insecurity on sub-Saharan Africa and to a lesser extent on South Asia. Within the developing world, the adverse impacts of climate change will fall disproportionately on the poor. Many quantitative assessments also show that the socio-economic environment in which climate change is likely to evolve is more important than the impacts that can be expected from the biophysical changes of climate change.

Less is known about the role of climate change for food stability and utilization, at least in quantitative terms. However, it is likely that differences in socio-economic development paths will also be the crucial determinant for food utilization in the long run and that they will be decisive for the ability to cope with problems of food instability, be they climate-related or caused by other factors.

Finally, all quantitative assessments we reviewed show that the first decades of the 21st century are expected to see low impacts of climate change, but also lower overall incomes and still a higher dependence on agriculture. During these first decades, the biophysical changes as such will be less pronounced but climate change will affect those particularly adversely that are still more dependent on agriculture and have lower overall incomes to cope with the impacts of climate change. By contrast, the second half of the century is expected to bring more severe biophysical impacts but also a greater ability to cope with them. The underlying assumption is that the general transition in the income formation away from agriculture towards non-agriculture will be successful.

How strong the impacts of climate change will be felt over all decades will crucially depend on the future policy environment for the poor. Freer trade can help to improve access to international supplies; investments in transportation and communication infrastructure will help provide secure and timely local deliveries; irrigation, a promotion of sustainable agricultural practices, and continued technological progress can play a crucial role in providing steady local and international supplies under climate change.

Bibliography

Baran, Paul A.: *The Political Economy of Growth,* New York, Modern Reader Paperbacks, 1975.

Basu, Kaushik. *Analytical Development Economics, The Less Developed Economy Revisited,* Cambridge, Mass.: MIT Press, 1997.

Brandwein, P.F.: *Sourcebook for the Biological Sciences,* San Diego, Harcourt Brace JOvanovich, 1986.

Caplin, A. and Schotter, A.: *The Foundations of Positive and Normative Economics,* New York, Oxford University Press, 2008.

Carl F. : *Working with Nature: Resource Management for Sustainability.* Australia: Harwood Academic Publishers, 2001.

Cootner, Paul : *The Random Character of Stock Market Prices,* MIT Press, 1964.

Dabholkar, A.R. : *General Plant Breeding,* Concept, Delhi, 2006.

Daniel Kahneman, Amos Tversky.: *Choice, Values, Frames,* The Cambridge University Press, 2000.

David Sadava: *Plants, Genes and Crop Biotechnology,* Sudbury MA, Jones and Barlett Publishers, 2003.

Desmond, Glenn, and John A, Marcell, *Handbook of Small Business Valuation Formulas and Rules of Thumb,* Valuation Press, 1993.

Dharamvir Hota: *Modern Biotechnology in Plant Breeding,* Gene Tech Books, Delhi, 2007.

Dodds, John H.: *Plant Genetic Engineering,* New York, Cambridge University Press, 1985.

Edwards, C.A. *Sustainable Agricultural Systems.* 1990. Ankeny, Iowa: Soil and Water Conservation Society.

Fraser, Steven, *Labor Will Rule, Sidney Hillman and the Rise of American Labor,* Reprint ed. Ithaca, N.Y.: Cornell University Press, 1993.

Fumento, Michael: *Bioevolution: How Biotechnology is Changing Our World,* San Francisco, Encounter Books, 2003.

Goleman D.: *Emotional Intelligence,* New York: Bantam, 1995.

Goodman, D.C.: *From Farming to Biotechnology: A Theory of Agro-industrial Development,* Oxford, Blackwell, 1987.

Gowdy, J., *Coevolution Economics: The Economy, Society and the Environment,* Kluwer, Boston, 1994.

Gregory H.; 1993. *Defining Sustainable Forestry.* Washington, D.C.: Island Press.

Guarti, Luigi, *The Valuation of Firms,* Blackwell Publishing, 1994.

Harrigan, K. R.: *Strategies for Declining Businesses.* Lexington, MA: Heath, 1980

Hausman, D. M.: *The Inexact and Separate Science of Economics*, Cambridge, Cambridge University Press, 1992.

Hunt, E. K. *History of Economic Thought, A Critical Perspective,* New York, HarperCollins, 1992.

Knapp, C. L.: *Commercial Damages, A Guide to Remedies in Business Litigation,* Matthew Bender, 1993.

Lal, R., : *Soil Quality and Agricultural Sustainability.* Chelsea, Mich.: Ann Arbor Press, 1998.

Link, Albert N.: *Evaluating Economic Damages, A Handbook for Attorneys*, Westport, CT, Quorum Books, 1992.

Murray, David: *Seeds of Concern: The Genetic Manipulation of Plants*, Sydney, University of New South Wales, 2003.

Peirson, G., Bird, R., Brown, R., & Howard, P.: *Business Finance*, Roseville, 1990

Pfeffer, J., and Salancik, G. R.: *The External Control of Organizations, A Resource Dependence Perspective*, New York, 1978.

Poindexter, J.S.: *Strategies and Mechanisms for Field Research in Environmental Bioremediation,* American Academy of Microbiology, Washington, D.C., 1994.

Schumpeter, J. A., *The Theory of Economic Development*, Harvard University Press,Cambridge, Massachusetts, 1934.

Scott, W. R.: *Organizations: Rational, Natural, and Open Systems*, Englewood Cliffs, 1981.

Selznick, P.: *Leadership in Administration, A Sociological Interpretation*, New York, Harper & Row, 1957.

Shetty, Kalidas: *Food Biotechnology*, New York, Dekker/CRC Press, 2005.

Smith, Gordon V., and Russell Parr, *Valuation of Intellectual Property and Intangible Assets,* New York: Wiley, 1994.

Steven E. : *Remote Sensing for Sustainable Forest Management.* Boca Raton, Fla.: Lewis, 2001.

Tietenberg, Tom, *Environmental Economics and Policy,* New York, HarperCollins, 1994

Tinbergen, Jan.: *On the Theory of Economic Policy*; Amsterdam, North-Holland 1952.

Tollefson, Chris, : *The Wealth of Forests: Markets, Regulation, and Sustainable Forestry.* Vancouver: UBC Press, 1998.

Trugman, Gary R., *Guide to Conducting a Valuation of a Closely Held Business*, New York, American Institute of Certified Public Accountants, 1993.

Weber, M.: *The Theory of Social and Economic Organizations*, New York, Oxford University Press, 1947.

Weibull, J. W.: *Advances in Understanding Strategic Behaviour*, New York, Palgrave, 2004.

West, Thomas L.: and Jeffrey D. Jones: *Handbook of Business Valuation*, New York: Wiley, 1992.

Woodward, J.: *The Oxford Handbook of Philosophy of Economics*, New York, Oxford University Press, 2009.

Yegge, Wilbur M., *A Basic Guide for Valuing a Company*, New York, Wiley, 1996.

Index